Library of
Davidson College

de Gruyter Studies in Organization 63
Experts in Organizations

de Gruyter Studies in Organization
Innovation, Technology, and Organization

This international and interdisciplinary book series from de Gruyter presents comprehensive research on the inter-relationship between organization and innovations, both technical and social.
It covers applied topics such as the organization of:
- R & D
- new product and process innovations
- social innovations, such as the development of new forms of work organization and corporate governance structure

and address topics of more general interest such as:
- the impact of technical change and other innovations on forms of organization of micro and macro levels
- the development of technology and other innovations under different organizational conditions at the levels both of the form and the economy.

The series are designed to stimulate and encourage the exchange of ideas between academic researchers, practitioners, and policy makers, though not all volumes address policy- or practitioner-oriented issues.
The volumes present conceptual schema as well as empirical studies and are of interest to students of business policy and organizational behaviour, to economists and sociologists, and to managers and administrators at firm and national level.

Editor:
Arthur Francis, Glasgow University Business School, Glasgow, GB

Advisory Board:
Prof. Claudio Ciborra, University of Trento, Italy
Dr. Mark Dodgson, The Australian National University, Canberra, Australia
Dr. Peter Grootings, CEDEFOP, Berlin, Germany
Prof. Laurie Larwood, Dean, College of Business Administration. University of Nevada, Reno, Nevada

Armand Hatchuel and Benoît Weil

Experts in Organizations

A Knowledge-Based Perspective
on Organizational Change

Translated by Liz Libbrecht

Walter de Gruyter · Berlin · New York 1995

France edition: Armand Hatchuel · Benoît Weil
L'Expert et le Système
Gestion des savoirs et métamorphose des acteurs dans l'entreprise industrielle
suivi de Quatre histoires de systèmes-experts
Economica, 1992
49 rue Héricart, 75015 Paris

With 9 figures and 2 tables

∞ Printed on acid-free paper which falls within the guidelines of the ANSI to ensure permanence and durability.

Library of Congress Cataloging-in-Publication Data

> Hatchuel, Armand.
>   [Expert et le système. English]
>   Experts in organizations : a knowledge-based perspective on organizational change / Armand Hatchuel and Benoît Weil ; translated by Liz Libbrecht.
>   — (De Gruyter studies in organization : innovation, technology, and organization ; 63)
>   Includes bibliographical references and index.
>   ISBN 3-11-014119-1 (alk. paper)
>   1. Expert systems (Computer science)—Industrial applications. 2. Production engineering—Automation. 3. Expert systems(Computer science)—Industrial applications—Case studies. I. Weil, Benoît. II. Title. III. Series: De Gruyter studies in organization ; 63.
> TS176.H37513 1955
> 658.5'14—dc20
>   95-22226
>   CIP

Die Deutsche Bibliothek — CIP-Einheitsaufnahme

> **Hatchuel, Armand:**
> Experts in organizations : a knowledge-based perspective on organizational change / Armand Hatchuel and Benoît Weil. Transl. by Liz Libbrecht. — Berlin ; New York : de Gruyter, 1995
>   (De Gruyter studies in organization ; 63)
>   ISBN 3-11-014119-1
> NE: Weil, Benoît:; GT

© Copyright 1995 by Walter de Gruyter & Co., D-10785 Berlin
All rights reserved, including those of translation into foreign languages. No part of this book may be reproduced or transmitted in any form or by any means, electronic or mechanical, including photocopy, recording, or any information storage and retrieval system, without permission in writing from the publisher. — Printed in Germany.
Converted by: Knipp Satz und Bild digital, Dortmund. — Printing: Gerike GmbH, Berlin. Binding: Dieter Mikolai, Berlin. — Cover Design: Johannes Rother, Berlin.

Contents

Acknowledgements		XI
Introduction		1
1	Expertise faced with diversity	3
2	Expertise as an object of management	4
3	Expert systems: a questionable development	6
4	Two ways of dealing with the same material	6

Part 1 9

Chapter 1
Exploring expertise
Objectives and materials of a study 11

1	Expert systems: the attributes of an approach	12
1.1	Elements of the method	13
1.2	A new management technique	14
2	Expert systems: a vehicle in the domain of expertise	16
3	Research material: expert systems in the industrial world	19
3.1	Naval: a paradoxical experience	19
3.2	Long-term follow-up of projects	20
3.3	A methodology for monitoring projects	21
3.4	Three basic hypotheses: rationalization with a participatory nature	23

Chapter 2
Artisan, repairer, strategist
Different facets of expertise 27

1	Knowledge and reasoning: where is the boundary?	27
1.1	A practical but limiting distinction	28
1.2	Classification of expert systems: getting back to the nature of expertise	29
2	"Doing know-how" or the artisan's expertise	30
2.1	The preparation of routings for metal processing	30
2.2	The representation of "doing know-how": a typical structure for expert systems	32

2.3	The role of calculations and technological models	34
3	"Understanding know-how" or the repairer's expertise	36
3.1	"Doing know-how" and "understanding know-how"	36
3.2	Cornélius: maintaining a flexible cell	37
3.3	Appropriate knowledge for diagnosis: the development of a suitable physiology	38
4	"Combining know-how" or the strategist's expertise	43
4.1	Naval: planning the use of oilrigs	44
4.2	The GESPI project: untangling the web in a large station	49
4.3	Technology and expertise	53
5	A paradigm for multiple expertise?	54

Chapter 3
Life of Expertise and Metamorphosis of actors
Birth, crises and development of expert systems 57

1	Birth of the project: myths and innovators	58
1.1	Mythologization of an industrial problem	59
1.2	Innovators: specialists or interveners?	62
2	Experts in organizations: nature of expertise and position of the actors	64
2.1	Workshop planners: a trade born with Taylorism	65
2.2	Station traffic planners: logic of the station	66
2.3	Maintenance specialists: organizing expertise hierarchically	67
2.4	Scheduling oilrigs: experts or negotiators?	68
3	Dynamics of the projects: multiple lines of transformation	69
3.1	TOTEM: the basis of a new engineering science	70
3.2	GESPI: from station traffic planner to network planner	72
3.3	Cornélius: the problem of transferring expertise	73
3.4	Naval: the loss of actors and their expertise	75
4	Organizational change: production of expertise and metamorphosis of actors	77
4.1	The limits of organization seen as a game	79
4.2	Production of expertise and control of the framework of collective action	82

Chapter 4
The nature of management techniques
Dynamics and unexpected repercussions of rationalization 85

1	Portrait of an expert system project: from myth to stakes	86
1.1	A mobilizing myth	86
1.2	Enhancing and sharing knowledge	87
1.3	Transformation of expertise and discovery of stakes	88

2	From operational research to expert systems: a new industrial logic	89
2.1	Operational research: modelling to optimize	90
2.2	Operational research and expert systems: different origins	92
2.3	Expert systems: operational research dedicated to new industrial logics	92
2.4	Forgotten lessons of operational research: the problematics of integration	94
3	On the nature of management techniques	95
3.1	"Rational myths"	96
3.2	A ternary structure	97
3.3	Use of a management technique: reacting to three types of incompletion	100
4	The AI actor: an autarchic producer of expertise	100
5	The unexpected repercussions of rationalization: the progressive discovery of collective action	102

Chapter 5
Hidden crises of industrial expertise
Practical and cultural stakes in expert systems ... 105

1	Industrial stakes in expert systems: the hidden crises of industrial expertise	106
1.1	Conditions for transferring expertise: relevance and common knowledge	106
1.2	The increasing complexity of industrial expertise: proliferation and heterogeneity	108
1.3	The effects of technological and economic imbalances: singular expertise	111
1.4	Mutation of planning and design expertise and management tools	112
2	Firms as producers and legitimizers of expertise	116
2.1	Training approaches: new relations between work and learning	117
2.2	Weakening of hierarchical relations and dynamics of expertise	118

Chapter 6
Conclusion ... 121

1	Problems of automating expertise	122
2	A more accurate perception of rationalization	123
3	The stakes of a better sharing of knowledge	123

Part 2
Four case histories of expert systems 127

Introduction 129

Chapter 1
TOTEM
The reconstruction of production planners' expertise 131

1	Problem and context: metallurgy and variety	131
1.1	Processing precious metals: consequences of diversity	131
1.2	The role of production planners and routing	132
1.3	Emergence of the project: the goals	133
2	Expertise in action: production planners prepare their routing	134
2.1	Example of a production routing	134
2.2	Preparation of a production routing: steps in the reasoning	135
2.3	Production routing: a complex task	137
3	The reconstruction of automated knowledge: birth of a new type of routing	139
3.1	TOTEM: an expert system for generating routings	139
3.2	Developing TOTEM: the procedure	143
3.3	Knowledge formalization: an instructive phase	145
3.4	Division between technical expertise and operational expertise ...	147
3.5	A judicious criterion for choosing applications: the existence of generalizable expertise	149
4	The project and the development of stakes: towards an industrial transition	150
4.1	Implementation and conditions for the integration of the system ..	150
4.2	First elements in the evaluation of a project	152
4.3	Emergence of a new type of methods engineer	154

Chapter 2
Cornélius
Fragmented expertise of maintenance specialists 155

1	Problem and context: controlling a facility	155
2	Expertise in action: the maintenance team	157
2.1	Operators of a highly automated flexible cell	157
2.2	A two-tier maintenance service	159
3	Reconstruction of automated knowledge: expertise and usage	159
3.1	Cornélius: an expert system dedicated to maintenance	160
3.2	Physiology of the machine: structuring knowledge bases	163
3.3	Experimentation and fine-tuning: who were the users to be?	164
3.4	Variety of uses, variety of knowledge	166
4	The project and the development of stakes: lost relevance	168

Chapter 3
GESPI
Discovery of station traffic planners' expertise 171

1	Problem and context: activity in a large railway station	171
1.1	The station: a railway traffic node	171
1.2	A problem which calls for formalization	172
2	Expertise in action: the station traffic planners	173
2.1	Three levels of traffic control	174
2.2	Limiting initial ambitions	175
2.3	The unseen staff in the planning department	175
3	Reconstruction of automated knowledge: progressive discovery of expertise	176
3.1	GESPI: knowledge and reasoning	176
3.2	The project underway	179
4	The project and development of stakes: rethinking the station?	186
4.1	GESPI in the hands of the planners	186
4.2	Evolution of the traffic planners' job	191
4.3	Multiple facets of project evaluation	192
4.4	The planning department at the centre of the station and its changes	195
4.5	Prospects for diffusing GESPI	196

Chapter 4
Naval
Undefinable expertise of strategic planners 199

1	Problem and context: oil exploration and instability	200
1.1	Drilling: decentralization, uncertainties and commercial negotiations	200
1.2	The stand-by crisis of 1982	202
2	Expertise in action: planning without a planner	202
2.1	Response to the stand-by crisis: strengthening co-ordination mechanisms	202
2.2	Planning with very few tools	204
3	Reconstruction of automated knowledge: imbalance of expertise	206
3.1	First steps of the project: who is the planning expert?	206
3.2	Naval: building a programme	208
3.3	A pilot for Naval: a system for experts	213
4	The project and the development of stakes: a new status for the experts?	214

4.1	Towards a central planning team	214
4.2	Effects of the counter-shock: the value of planning in a cyclic environment	215
4.3	End of the project: the loss of expertise	216

Bibliography 219

Acknowledgements

This book would not have been possible without the active and prolonged co-operation of the firms whose experiences with expert systems we studied. The analyses we propose were discussed in detail with the project leaders and other participants concerned. Together with them we validated the facts presented in this book. The choice of anonymity, wherever possible, was made by the authors; we felt that information on identities was out of place in a book of general interest.

The research on which this book is based was funded partly by the French Research and Technology Ministry and in particular its AMES programme.

The book also benefited from the constant interchange between the authors and their colleagues at the Centre de Gestion Scientifique of the Ecole des Mines de Paris.

The present English edition owes much to the advice of Mathew Liberatore (Villanova University), Louis-Georges Soler (Institut National de la Recherche Agronomique, Paris), Bengt Stymne (Stockholm School of Economics) and to the work of the CNRS Groupement de Recherche Relations Professionnelles.

The Centre National des Lettres of the French Ministry of Culture helped to finance the translation.

The authors would like to thank Ms Liz Libbrecht who, in her translation, was able to remain faithful to the spirit of the French text and to take into account the peculiarities of international literature in the fields dealt with.

Introduction

> Falsity consists in the privation of knowledge, which inadequate, that is to say, mutilated and confused ideas involve (Spinoza, *Ethics III*, prop. 35).

What will be the shape of firms in the twenty-first century? Will they be "inverted pyramids"[1] with only the simplest of hierarchical structures? Will tomorrow's organization charts describe, not services, but rather networks of actors transformed into just as many individual companies? Are firms likely to evolve in the same way as community movements, driven by strong collective cultures or charismatic leadership? The formulae proposed over the past few years by certain management theorists[2] clearly indicate a renewed quest for the ideal firm, the "one best way" to prefigure the efficient enterprise of tomorrow. This prevalent concern with a mobilizing doctrine is all the more striking in that it contrasts with the main trend in the 1970s. At the time, the virtues of a certain relativism were put forth and the multiplicity of organizational forms was perceived as an adequate response to the diversity of situations. However, renewed interest in the "right model" should be taken seriously, for it may well be a symptom of the conceptual and practical difficulties experienced by many firms faced with the profound upheavals that have affected economic life during the past twenty years. The perpetual slump, the arrival of new competitors, or the instability of markets have eroded many certainties and spawned a number of new doctrines. This need for a model is understandable when transformations with multiple and uncertain impacts have to be undertaken in a threatening and difficult environment. It provides guidelines and can at least be compared to the contrasting model that is henceforth, and sometimes too habitually, constituted by the so-called Taylorian or bureaucratic firm.

In imagining the evolution of firms or that of their operation we are perhaps as much at a loss as an observer in the 1880s trying to forecast the shape of the twentieth century economy. Having witnessed, like us, a long economic slump, a proliferation of new technologies (including the automobile and electricity) and the emergence of new powers (Germany and the United States), he or she would probably have imagined neither Taylorism, nor the monthly payment of salaries, nor marketing departments. In many respects the transformations in economic life that we are witnessing today are a reminder, in their intensity and their extent (proportionately speaking), of those at the end of the nineteenth century. We must therefore agree that an exercise in forecasting is somewhat futile, in spite of our convic-

---

1 In J. Carlzon's words (1986).
2 This movement was initiated by Peters and Waterman's book on the most efficient American firms. In France the same theme is found in the work of Archier and Sérieyx (1983). A synthesis of these different doctrines is presented by Aktouf O. (1989).

tion that future upheavals may well be even more profound than those which we have already experienced and that we must be prepared for them. These remarks leave few options for thought; *in particular they are an invitation to understand those trends which in current situations generate new questions and tensions, and those which are only emerging, by outlining the dominant questions of tomorrow's firms.*

Rather than finding the model of an ideal firm (and we cannot say in what way it would be more efficient or better), we need to detect the problems that will convey the most important stakes and to which future firms, in order to be viable, will have to respond. *It is around these problems that the improvements to be made, the resources to be used and the organizations to be invented, will be defined.* Firms will of course always have to deal with decision making or financial issues, they will always have to choose and to motivate their employees in one way or another and to wonder about whether they should apply rules or give autonomy to the different actors, but the permanent nature of these problems hides a more essential reality. Each period and each context gives particular weight and content to one or another of these questions.

The development at the beginning of the 1980s of various forms of worker participation and involvement (e.g. progress groups and quality circles) can therefore hardly be explained if we think only in terms of employers' failure up until then to take an interest in their workers' abilities. It is only by looking at the new problems and specific historical context which made this participation relevant, that one can say whether there really was what we call *rationalization*, in other words, increased efficiency or – what amounts to the same thing – a solution for some organizational crises. Worker involvement was not merely a response to an intensified need for autonomy felt throughout society; it was also a means of counteracting the *compartmentalization of expertise* that had become dangerous in an economy oriented towards permanent innovation and the proliferation of products. For if we lived in an industrial world that was perfectly controlled by its designers or that incessantly repeated the same routines, what would be the use of motivating its operators to carry out a permanent search for defects and dysfunctions. What would be the use of the urgent demand for their initiatives? Behind these sometimes spectacular new forms of work lies the emergence of a series of problems concerning not only the structure of industrial firms or their decision making processes but also, more profoundly, the dynamics of the expertise which keeps economic entities alive.

The research presented here, although devoted to the development of expert systems in industrial firms, is the outcome of such analysis and therefore of *the hypothesis that, in an economy where product variety and innovation are vital, expertise necessarily constitutes a new and favoured field of rationalization.*

1 Expertise faced with diversity

The question of expertise itself is by no means new; firms are constantly preoccupied with procuring required skills through recruitment[3] or training. However, the significance of these problems has become apparent on a global scale where it involves national industrial and technological policies[4]. On a corporate level it naturally involves the stakes linked to research and development, the control of strategic processes, and the problems raised by the accelerated renewal of products and markets. The latter might only have had limited consequences, had this renewal not been accompanied by the increasing complexity of goods and services in the most advanced economies. The number of optional extras and variations of manufactured products has become infinite and it is becoming clear that a "good product" cannot be identified by its price, qualities, image, after-sales service or delivery lead time alone, but by a variable combination of all these attributes. In order to define current issues, the two notions of renewal and complexity must therefore be inseparably linked; in other words, the name of the game must be "variety economy"[5], with this term taken in its fullest sense, i.e. meaning the multiplication of types and the constant creation of new variations of products in any firm.

What then are the consequences of such complexity and how do firms react to it? Besides the employee involvement mentioned above, the current emphasis on training is a significant type of response to this destabilization. The number of innovative experiments in this field has multiplied, which also points to the importance of the stakes and probably – according to recent research – to the creation of new expectations in employees and management alike[6].

A different type of approach was initiated concurrently in the 1980s with the development of expert systems. The innovation was a particularly ambitious one in so far as it aimed for improved control and diffusion of knowledge, and in that it

---

3 The international circulation of artists or craftsmen is as old as civilization itself. Examples akin to our current preoccupations can easily be found in the early nineteenth century when companies turned to England for specialists in the construction of industrial machines.
4 On this subject see Dosi G. et al. (1988); Cohen E. (1989).
5 The concept of variety is narrowly linked to the more common concept of flexibility; however, in this work variety means essentially product customization and innovation and can be grossly evaluated by the number of different final articles produced by the same production system; see Cohendet A. et al. (1988); Hatchuel (1988a).
6 See Riboud A. (1987); Dubar (1991). Spurred by the Japanese model, the movement towards work reform and new shop management was intense during the eighties. However, it focussed on shop workers and rarely analyzed the technical departments. The members of these departments will, by contrast, be the heroes of the expert system projects studied in this book.

inevitably implied the *institution of new relations between the experts themselves and the system in which their expertise was used.*

Before looking more closely at this approach, it may be useful to clarify the meaning given to the word *expertise*. Bearing in mind our general computer culture, it is important to specify that expertise is not an information system or a database and *that it comprises a set of theories and questions on which an activity can be based, or from which data can acquire meaning by generating new theories or questions.* Nor is expertise, by nature, a discipline or a science: it has not necessarily been developed systematically or subjected to academic control procedures; it can be constituted in a variety of ways and base its legitimacy on complex mechanisms (to be discussed below). Take, for example, a salesman who visits his clients in a particular region fairly regularly. He gathers facts, develops certainties, and constantly retains a series of doubts and questions. This living body of knowledge and questions is essential to his activity. Even if it is riddled with mistakes and short-cuts, or combines a number of facts, opinions, calculations and beliefs, it will still function for him, or for others, as expertise, in other words as a source of references and ideas, or an instrument of action and communication[7]. In corporate life, new expertise is constantly being created through the entire range of possible activities, from material processes to commercial or legal exchange, interpersonal relations or modes of organization. In short, the question of expertise in firms is clearly not limited to the fields generally referred to as technological.

2 Expertise as an object of management

Towards the end of the 1970s, artificial intelligence (AI) and more particularly its most prevalent branch, expert systems (ES), became a leading new technology that nourished numerous philosophical debates and heralded a new computer revolution. But above all, through it the idea of using computers to gather expertise on any subject domain became feasible. During the first half of the next decade, leading firms soon set the trend – some even boasted about their use of expert systems in their advertising messages – and a series of projects and experiments was launched. It was not that those firms which decided to build expert systems had suddenly discovered that their employees had *expertise*, but rather that it had suddenly become possible to preserve, update and distribute this expertise by means of computers. *Expert systems thus opened a new field of knowledge and action:*

---

7 This type of definition also helps to avoid the natural tendency of implicitly confusing knowledge and techniques and, more specifically, knowledge and techniques relating to matter or machines.

they turned expertise into the object of possible rationalization, into an object of management, in the same way, for example, that Taylorism had made of time and motion an object of investigation and control.

Were expert systems the best approach for rationalizing expertise? The limits of the project were indeed easy to define: who would have thought that the complexity of human knowledge or the learning capacities of homo sapiens could be found in a machine, even the most powerful? But this kind of limit has no sense when applied to corporate life[8], an expert system containing even a small portion of knowledge may be economically adequate. No judgement could therefore be passed on the approach without a preliminary understanding of the dynamics of expertise in industrial firms, or without an idea of the possible changes and problems to which the restructuring and redistribution of this expertise might lead.

As surprising as it may seem, these questions did not play a significant part in reflection on corporate functioning. Research most often focussed on problems of structure, strategy and more recently culture; the study of actors' expertise was however spurred by problems concerning technological change[9].

Admittedly, the subject does not lend itself very readily to investigation or observation, for a person's expertise is not something that is easily accessible. Although clues can be picked up in certain types of behaviour, the main elements constituting expertise may still escape analysis. *Expert systems, on the other hand, forced interested firms to undertake fundamental research on the knowledge held by some of their main actors. We therefore felt that this type of project, as much through its success as through its difficulties, provided a perfect opportunity for exploring the dynamics of expertise in organizations, precisely because it embodied a more extreme form of rationalization.*

---

8   From the beginning of the 1970s Hubert Dreyfus (Dreyfus H.L., Dreyfus S.E. 1986) attempted to demonstrate the limits of artificial intelligence. The debate initiated by him is not, however, directly relevant here, for knowing whether AI programmes will effectively culminate in a machine that "thinks", bears very little relation to the potentially industrial dimension of these approaches. On the other hand, these philosophical debates contributed to the "futurist" image of expert systems.

9   Mention must nevertheless be made of comparative studies on educational systems which looked at the impact of qualification systems on the organization of firms, even if they did not specify what form of knowledge is produced by these systems (Maurice M. et al. 1986).

3 Expert systems: a questionable development

In itself this conclusion might have led to the belief that expert system projects could provide instruments for research on expertise in firms, had there not been so many practical questions which remained unanswered by manuals and specialized reviews. What happened to the basic assumptions of expert systems in the field? At what expertise was the approach aimed? How could expertise be gathered in an organization? Since the advent of expert systems in the 1980s, certain specialized firms have clearly lost interest in them and a large number of projects have not advanced beyond the planning stages. Only very recently has an improvement in this situation been recorded by specialists in the field. There was thus a convergence of our interest in understanding the stakes involved in the rationalization of expertise, and the problems encountered by firms undertaking such projects. *It was this twofold interest which guided our research and the structure of this book.*

Most of the material used was taken directly from real projects whose development we monitored on an on-going basis. Four of these have been described and discussed in detail in Part Two; *they concern the automation of production routing in a precious metal industry, breakdown diagnosis of machining equipment in an advanced mechanical firm, the assignment of routes to trains in a large railway station, and finally the planning of exploration in an oil company.* The four projects were chosen amongst several others because they mobilized different types of knowledge and actors with differing status, but also because there was a genuine intention to make them operational. The study of their development, their crises, their options, and the organizational transformations associated with them, formed the basis of our analyses of the management of expertise in industrial firms – which also explains why this book comprises two main parts.

4 Two ways of dealing with the same material

Part One is devoted to the presentation of analyses and conclusions drawn from our research. The different projects are discussed concurrently, along a number of essential themes.

Part Two describes each project separately, along the main lines of a common thematic structure. It is not an annex, but is complementary to the first part and constitutes a more narrative and extensive presentation of the same material. Our intention was to offer readers direct and overall access to the projects studied, thereby allowing them to draw conclusions that differ from those based on our own assumptions, suggested in Part One. It is also a response to the need which we encountered in many of our interlocutors, for greater detail on the technical, economic or social aspects of such projects. Such detail would have made the first part somewhat unwieldy, had it been included there.

To conclude this introduction with an overall view of the main arguments presented, we shall briefly summarize *the contents of the different chapters comprising the first part of the book.*

Chapter 1 presents the basic hypotheses on which expert systems are founded, as well as the methodology and analysis framework that we used.

Chapter 2 characterizes the expertise which served to develop the systems in the different projects studied. We see how the basic hypotheses of the approach had to be transgressed, because they described a very limiting form of expertise. Thus, although expert systems are defined by the imitation and harnessing of human knowledge, the success of projects necessitates the transformation, even the enhancement, of that knowledge. This restructuring of expertise can be the condition for automation, and is in itself already a process of rationalization. The transgression of the basic hypotheses of ES demonstrates the diversity of forms of expertise in action, diversity which we tried to condense by distinguishing three main types: that of the artisan, the repairer and the strategist.

Chapter 3 shows how the restructuring of expertise, which takes place during the course of the project, transforms relations between actors. This type of change goes beyond a mere modification of the powers or respective responsibilities of individuals; it leads, in certain cases, to what we shall call a metamorphosis of the actors, going as far as the birth or disappearance of some of them. The transformation of expertise implies a potential displacement not only of means for action, but also of the very definition of actors. It does not therefore amount to a mere game between predetermined actors. Some of its theoretical characteristics are outlined in this chapter.

Chapter 4 adopts a more historical perspective. We see that expert systems extend the rationalization process started at the end of the last century, whilst simultaneously renewing its objectives and approaches. They help us to better understand the nature of *management techniques,* in which they constitute the latest known trend. We show how each of these is composed of three structural elements: a technical substratum, a managerial philosophy and a simplified organizational model. But the scientific organization of work, operational research, new forms of flow management or expert systems, all make over-simplified assumptions on the conditions of collective action and must therefore be called into question. This counter-culture, because it illuminates the misunderstood aspects of corporate life, is an effective vehicle for efficiency and constitutes an integral part of the rationalization process.

Chapter 5 looks at the main industrial stakes revealed by expert system projects. *Production planners, station traffic planners or maintenance specialists whose expertise was elicited in the projects studied, are all products of Taylorism.* Artisans of its implementation and supported by it, they became part of industrial firms at the start of the century, and there they invented the figure of a new actor who based his legitimacy on specific know-how and the position of a conceiver. The study of their expertise shows them caught between the figure of an engineer and that of a negotiator; it also reveals the heterogeneity of their knowledge, aggravated by the consequences of a variety economy found in all the situations studied. It is when living expertise, generated by uncontrolled accumulation by a small number of actors, has to be restructured that the main stakes of rationalization appear most clearly. Expert system projects thus help to reveal the hidden crises which may await these actors. They show us to what extent industrial firms are themselves faced with a crisis in their central expertise, since this expertise has to be restructured to respond to the complexity of the economic environment.

Like its predecessors, the expert system technology gives rise to a type of unexpected repercussion. It reminds us that organizations must be considered as systems of knowledge production – *a view that will be all the more useful if we are to experience the weakening of hierarchical or bureaucratic structures, in favour of more participatory horizontal processes. In order to understand new organizational forms, the knowledge systems of actors and the interaction which these allow must be taken into account.*

It is therefore hardly surprising that we have to revise our conception of corporate life if we want to understand the effects and dead ends of rationalization.

Part 1

Chapter 1
Exploring expertise
Objectives and materials of a study

We shall start with an imaginary outing and a little riddle. Invited by an industrial firm which could be qualified as modern, we are taken on a guided tour through a series of departments, ranging from planning to maintenance. Since industrial modernity is often – and usually rightly so – linked to computerization, it seems quite normal for us to see a number of people with their eyes glued to computer screens. Three of them are of particular interest to us, although nothing obvious distinguishes them from their colleagues. One, we are told (although we do not know who he is), uses a special kind of advanced, new generation computing system that is "intelligent" – an "expert system"!

Surprised and intrigued, we want to know more. The first person explains that he has to set his machine before starting it. By answering a list of questions proposed on the screen, he can define the desired characteristics of the part to be manufactured. In return the computer will calculate and provide the most suitable setting for the machine. The second person seems to operate with a broader perspective. He is busy with his weekly task of checking the overall production programme for the workshop, proposed by his production management software. In order to draw up this plan, the computer takes into account a large body of information from the sales department, research department, or workshops, which means that its proposals cannot easily be refused. However, the chance of it making a major error cannot be discounted, and the operator considers it wise to check the plan before approving it. The third person in our fictitious visit is clearly not as calm as his colleagues. The behaviour of one of the main facilities in the workshop has become a cause for concern; it is obvious that something is wrong, but what? Our man has some ideas, but needs confirmation, and he impatiently answers the multiple questions put to him by the computer program specially designed for diagnosing machine failures.

Would it possible, after such a visit, to say which of the three applications included the most "artificial intelligence"? Might one have the impression that diagnosing a failure required "expertise", whereas setting a machine or drawing up a production programme did not? Of course not!

1 Expert systems: the attributes of an approach

Yet this is precisely the kind of judgement that the notions of artificial intelligence and expert systems (of which failure diagnosis is the most commonly referred-to application, although not the easiest to implement) have inspired during the past ten years or so. These notions have profoundly, but ambiguously, influenced our way of viewing the computerization of firms, starting with the most common terms that ordinary users have had to assimilate. The classical computer model which distinguishes "programs" and "data", calculations and variables, has been replaced by the more complex structure of expert systems. Several new concepts have appeared, notably "knowledge bases", "inference engines" and "fact bases". Before defining them briefly, we must point out that these terms could well have represented mere refinements to existing computer language, or variations in the art of programming; had this been the case, expert systems and a good deal of artificial intelligence would never have had the repercussions we know them to have had these past few years. For beyond their technical contents, these expressions evoked from the outset a vast project and many problems which, carried by far more metaphorical meanings, led expert systems to rapid renown that exceeded the milieu of computer specialists.

Thus the definition usually given to expert systems, even in specialized manuals, immediately plunges one into imagery and allusion. They are described as computer programs aimed at "imitating human reasoning", and whose development therefore necessitates the "extraction of expertise" from experts in a particular subject domain. It is the combination of a set of more or less recent computer techniques which use abstract logical formalism on the one hand, and a less specific discourse on knowledge, know-how or expertise on the other hand, which forms the peculiarity of the field of action progressively created by means of AI and expert systems. Without this combination, the concepts of "knowledge representation" or "imitation of human reasoning" would soon have lost their specificity and could have been claimed by any computer program, since they all involve some degree of knowledge. The three examples in our imaginary visit all require complex knowledge or expertise before being implemented, even if in practice two of them can function without the use of AI tools.

It was thus by going beyond common sense, with a narrow and formal approach to "knowledge" and "reasoning", that expert systems were able to inaugurate a new field of thought and open the way to projects hitherto considered as unfeasible.

1 Expert systems: the attributes of an approach

1.1 Elements of the method

Two specific principles of method characterize expert systems. We shall get back to these again later, but at this stage they can be defined generally by means of a very simple example[1].

The first of these principles lies in the extensive use of deductive or inductive reasoning, essentially of syllogisms, which operate on elementary knowledge and make it possible to base reasoning on qualitative statements. It is these statements which specialists call "declarative" knowledge, as opposed to the equations and variables that prevail in traditional computing. To illustrate this idea, consider a very limited expert system which at first has nothing in its knowledge base other than the statement "All humans are mortal". If we inform the system that "Socrates is human" (a new fact) and if it is equipped with elementary rules of logic, it will conclude that "Socrates is mortal". We can thus see how our little expert system utilized its limited knowledge to produce a new assertion. If, however, we then tell it, not that "Socrates is human", but that "Socrates is mortal", and then ask "Is Socrates human?", the system has to solve the problem by trying to find out whether the knowledge it has allows it to reach this conclusion logically. In this case it would not of course be able to do so, for there is no basis to conclude that Socrates is a man. Since the expert system only knows that he is mortal and that all men are mortal, Socrates could well be a dog!

To continue this exercise, we would have to enhance our system's knowledge base by adding, for example, that "All philosophers are human". In trying to answer the question "Is Socrates a man", our system will check whether there are facts which, acting as a premise, allow such a conclusion. It will then notice that the statement "Socrates is a philosopher" is such a premise, but is it true? Has this fact been confirmed or else invalidated and, if there is still some uncertainty, can it have this removed by the user? The system must then stop and ask its user the question "Is Socrates a philosopher?". A dialogue between the user and his system is thus started, with the latter constantly trying to solve a problem by means of the knowledge at its disposal, occasionally asking for confirmation of facts which allow it to progress. If the answer to the question "Is Socrates a philosopher" is "yes", the expert system will finally state: "Socrates is human", and may also conclude that "All philosophers are mortal". If the answer is "no", it would of course state its inability to solve the given problem.

This example, despite its extreme simplicity, illustrates most of the basic concepts which were used to develop expert systems.

Knowledge bases generally consist of statements of the type "if A then B" or "all As are B". They can also contain the description of objects characterized by attributes or by relationships with other objects. But what happens with more com-

---

[1] For a more in-depth approach see : Harmon P., King D. (1985); Farreny H. (1986); Gondran M. (1986); Van der Gaag L., Lucas P. (1990).

plicated problems such as "From what illness is this patient suffering?" or "What wine should be served with this meal?"? The system then has to process a far greater amount of knowledge and very long reasoning strings which comply with more or less elementary rules of logic. Building expert systems capable of answering such questions therefore implies that the computer can accept knowledge in the form of statements in natural language, that it can organize and process the accumulation of statements which comprise, for example, medical or gastronomical know-how, and that it can also make rapid and suitable deductions and inductions enabling it to utilize its knowledge to answer the questions put to it.

These elements explain the origin of the second principle of method peculiar to expert systems. *This consists of the analysis and construction of programs, with a systematic separation between the structuring of knowledge and that of reasoning.* If the required reasoning all amounts to logical chains that use universal rules of logic, it becomes possible to set up standard tools devoted to this function and utilizable as such in any subject domain. These tools are the "inference engines". They often form the most technical part of the expert system, the black box, in which the statements contained in the "knowledge bases" are processed.

These, then, are the main attributes of expert systems. Each of them has, of course, been technically refined in numerous ways and we have merely outlined the rudiments of the discipline. However, they suffice for anybody wanting to understand the most common projects in practice, and they made it possible to launch this technique and its philosophy in the business world. With features and assumptions such as these, hopes could be placed very high, notably for the capturing of human "expertise" and the use of computers to solve problems that could hitherto be handled by experts alone. The scope for such a project seemed infinite, provided the right languages or computer tools to ensure speed and user-friendliness were available. It was at the start of the 1980s that these tools could be found on the market, accessible to firms of all sizes; the "expert system wave" was ready to spread.

1.2 A new management technique

Yet, if we consider expert systems from the viewpoint of corporate history and isolate their main characteristics, it is clear that despite their unquestionable originality, *there is something distinctly familiar about this type of approach.* Are expert systems not to knowledge what former movements had attempted to be to other objects fraught with economic implications? (For example, the achievements of operational research in the 1950s to improve decision making or, even well before that, the attempt to increase the productivity of manual labour by means of time and motion studies, the spearhead of Taylorism.) Such a perspective has more than a purely historical aim; it precludes the type of amnesia that usually accompanies innovation and that has also accompanied expert systems, and raises a number of questions that are likely to play an essential role in their development. As the chronometric measurement of tasks in factories at the beginning of the century cannot be reduced to the mere sophistication of the measurement of time, so the possible significance of expert systems, in other words, their capacity to provide a medium for a new management philosophy, should be considered carefully. Thus, one could have foreseen the necessity to *show every time how certain actors' knowledge should be considered as a resource or capital*, and how a better utilization of this knowledge could be realizable and weighted with economic or organizational stakes. That alone is what can be of real consequence to firms, and make the development of such systems an axis for evolution or reconstruction.

In so doing, expert systems have breathed new life into the old but dormant project of controlling intellectual tasks. With the concepts and tools making it possible to formalize knowledge, *they have invented a new form of rationalizing industrial life, and enhanced the family of what can be called management techniques*. We use this expression to include in the same concept the time-measuring techniques so dear to Taylorians, as well as decisional calculations, expert systems and, more generally, all the formalized models relative to any particular aspect of corporate life. The more classical notion of administrative techniques would have been just as suitable had it not been so exclusively associated with the instruments of financial and economic analysis. This qualification is not only for the sake of classification, for it alerts us to the fact that expert systems, because they belong to this rather strange family, cannot escape from all the difficulties, ambiguities and controversies that corporate management techniques have spawned over the past century or more, that is, since the management and organization of firms have adopted a methodological and professional approach[2].

It thus seems quite likely that a part of the problems encountered in the development of expert systems has, *mutatis mutandis*, already been experienced by other management techniques, and dealt with more or less successfully. Studying expert systems and considering their contribution, their assumptions about human knowledge, or the fate of projects which claim to adopt this approach, therefore forces one to reconsider old but pertinent questions. It becomes necessary to take a new look at the "nature" of management techniques and at the lessons which can be learned from their introduction into firms.

---

2   Chandler A.D. (1977).

2 Expert systems: a vehicle in the domain of expertise

Management techniques are filled with traps for the unwary. A century after their birth, Taylorism and scientific management appear to researchers and historians as being far more varied and complex than management manuals or the sociology of organizations usually make them out to be[3]. The same applies to the other forms of rationalization that succeeded the first Taylorian principles. Only a few years ago, operational research and the decision making disciplines were also the subject of a number of controversies[4]. More recently, the methods of computer-assisted production control[5] or industrial logistics from Japan firmly resisted any hasty evaluation and always obliged analysts to consider the organization of industrial firms in depth.

One of the main reasons for these difficulties results from the opinion that management techniques are tools whose efficiency and influence can be gauged in advance. But the concepts dealt with by these techniques are too broad for a general evaluation. Moreover, such an opinion maintains useless theoretical borders. There is, on the one hand, a notion of organizations in which structures, actors, and relationships of co-operation or of conflict are to be found and, on the other hand, completely separately, a notion of managerial efficiency and technical rationality which depends on the above-mentioned tools. This separation forms the basis of theories which are both contradictory and recurrent. Since the beginning of the century, the same critiques of rationalization procedures have been put forward time and again. In particular, criticism of Taylorism or, more recently, of the technocratic tendencies of management too preoccupied with procedures and figures, is well-known. Such accusations are by no means groundless; any rationalization constitutes, to a greater or lesser extent, a Procrustean bed for the phenomena to which it applies. There is a natural tendency to lop off anything that resists the approach, while that which is neglected may well be vital. But even if criticism is fair, it cannot apply to the actual principle of the approach. Any representation, whatever its object or its basis, runs the same risk, and the Procrustean bed need not be rectangular, nor very solid, for it to retain its dangerous temptations.

This dialectic of the actor and his instruments, an old philosophical trap, readily gives rise to the most contradictory arguments. Thus, management techniques are considered to have played a significant role in the development of enterprise and the industrial world. Economic actors have on many occasions borrowed meth-

---

3 De Montmollin M., Pastre O. (1984).
4 These debates took place with different perspectives in Europe and America. An overview of the discussions in the French context can be found in Heurgon E. (1978).
5 Hatchuel A., Sardas J.C. (1990).

ods or principles from countries or organizations reputed for their efficiency, with the Japanese case being the finest example of this phenomenon. It is always this same argument, although demonstrated by means of negative examples, that is developed by those who emphasize the dangers of management techniques or their influence[6] on human behavioural patterns in organizations. For saying that a tool can be misleading, is implicitly extolling the right tool, or at least the one which has no appreciable effect or... the effect of a mere placebo! In contrast, other authors have attempted to demonstrate that management techniques, whatever their qualities, have in fact been of little consequence in comparison with the strategic incentive of the actors themselves, with their natural ability to divert any formal attempt to influence them[7]; or else that these techniques could not possibly offset the absence of a social climate to promote the motivation and involvement of the actors[8].

The contrast in these viewpoints bears witness to the difficulties of defining the "mode of existence"[9] peculiar to management techniques, more than a century after their identification as elements of economic life. The clearest and simplest sign of these difficulties is certainly to be found in the widespread tendency to evaluate management techniques in terms of success or failure, to ask "Does it work?", as if they were mere gadgets, and as if the answer to these questions lay in the definition of the technique. Accepted as natural, this attitude is not devoid of economic and social consequences. When they first appear new "management" techniques always become a craze, which is often followed a few years later by circumspection or even disillusionment. The survival of the firms concerned is nevertheless approached with more selectiveness and caution and is subject to a certain metamorphosis or learning procedure. The latter are, in the final analysis, the elements that best reveal the processes of change that actually took place.

Expert systems provide an excellent example. Today, certain professionals admit to adopting such a cautious attitude that they advance "under cover", using AI without making it known. The phenomenon has several causes. To understand it, one has to retrace the history of the actors that led expert system projects in firms, the specialized companies that set themselves up on this market, and the university courses that trained the specialists in the field. Our main interest here is less in this history itself, than in its repercussions in corporate life. We shall therefore attempt to discover the real impact of expert systems, as well as the reasons behind corporate decisions to use certain technologies and innovations.

---

6 Berry M. (1985).
7 Brunsson N. (1985); Pavé F. (1989).
8 Aktouf O. (1989).
9 We have borrowed this expression from the French philosopher Simondon G. who developed one of the most profound analyses of technology in his book *Du mode d'existence des objets techniques* (1969).

Our methodological approach is based on the view that expert systems are a vehicle for the study of the dynamics of expertise in firms, and draws upon a renewed theory of the role of rational models in practice[10]. In this conception, the process of rationalization is no longer a gateway to the "one best way" entirely known in advance and supposed to be the prerogative of certain actors, or independent of the historical context. *It constitutes the starting point and the vehicle of a model of intervention and innovation in a firm.* The creative value of such a model is in no way systematic or inevitable; it is never a cure-all. This virtue, when it exists, lies in the fact that the process of intervention is a source of new expertise, and that it can reduce certain internal and external tension in the firm, or create more fruitful tension[11]. *The effectiveness of such a process is therefore not inherent from the start, it is discovered progressively as fertile collective learning processes are set up, in other words, new shared expertise and new relationships.* Hence, the rationalization process is only creative if it contributes towards the concurrent invention of new conceptions and new social relations that are better adapted to a historical context and more acceptable to its actors. However, no strict rule for success will be invoked here; the creative nature of such a process can be recognized only in the framework of a social system whose rules and lifestyles define that which the actors involved perceive as a creation of wealth (for customers or users), or as acceptable relationships (in professional life or commercial exchange). These conditions also show the risks involved in any rationalization process which can generate inadequate conceptions and lead to the decline or death of a group; in any case, its validity remains dependent on a context that can change.

Our viewpoint is also soundly supported by recent work on innovation[12]. The latter, in spite of being focussed on more classical technical objects (scientific innovations), has the particular merit of rejecting the idea that innovations merely "diffuse" in society and proposing the idea of a long process during which the contents of the innovation change as much as the society that receives it.

Thus, it seemed to us that the development of expert systems provided the means for a detailed examination of expertise in firms. Such an investigation seemed particularly relevant not only because we, and others, supposed that expertise plays an essential and new role in the economy, but also and paradoxically, because it is one of the aspects that receives little attention from prevailing organizational theories. A discussion on this point may, however, appear too academic without a preliminary presentation of the projects which served as a basis for our investigation; *an investigation which was open-ended, for nothing was predeter-*

---

10 Our viewpoint is supported by several studies which, focussing on operational research and decision support, introduced a more dynamic and relational conception of rationality and therefore of the place and role of management techniques: Hatchuel A., Molet H. (1986); Moisdon J.C. (1985); Roy (1990); Ponssard J.P., Tanguy H. (1993).
11 Pettigrew (1990); Hatchuel A. (1988a).
12 Of which a synthesis can be found in Callon (1986).

mined, neither the fate of the definition and hypotheses of expert systems, nor the possible transformations in the firms which launched these projects.

3 Research material: expert systems in the industrial world

Our interest in expert systems dates back to the early eighties, and was spurred by the innovative character of the approach and the fact that is was often proposed for long-standing industrial problems such as workshop planning that had already been dealt with by operational research or decision support systems. We knew what the conditions of such an approach were: drawing up plans also means making compromises and creating the collective relationships which give these plans their relevance and effectiveness. In what way did the new approach promote the consideration of these dimensions?

3.1 Naval: a paradoxical experience

We were not alone; the same question was asked by certain managers of the oil company which had included us in their experiment on the use of expert systems for planning. The project, called Naval, lasted from 1985 to 1987 and was aimed at developing an expert system to plan the use of off-shore oil rigs leased by the company's various subsidiaries. Based on a description of available rigs and prospects, the system proposed a compromise plan which took into account multiple and heterogeneous constraints. In 1987, in spite of fairly satisfactory technical results, the project was voluntarily discontinued for several reasons, including the effects of the oil shock at the end of 1986. The plummeting crude oil price lost the expert system part of its *raison d'être*. This was compounded by another unexpected problem: there was nobody in the firm who could fulfill an organizational role corresponding to the knowledge and objectives of the system (we shall discuss this further at a later stage). In its death, Naval highlighted the complexity of the dynamics of expertise in an organized game.

To broaden our analysis we undertook a study, lasting at least two years, of other industrial projects which necessarily mobilized different types of expertise and were backed by a sincere ambition to achieve operational results.

3.2 Long-term follow-up of projects

Our first preoccupation was to explore the validity of expert system hypotheses in different contexts; it corresponded to our main research objective, focussed on the dynamics of expertise in firms.

The second consideration was to avoid projects which never developed further than the stage of a secretly tested model; there is always a profusion of such projects in the early years of development of a new approach. These models are usually just to "see what it's all about", without there being any genuine intention to implement them. Few actors are involved and the project is still-born, without much having been learned from it. It is, however, not easy to avoid this type of situation when one wants to follow the actual life of the project and avoid, if possible, a retrospective study. Only those projects for which there are sound reasons to believe in their continuation, should be chosen.

That is partly why we decided to limit ourselves to conventional industrial contexts. Although the majority of projects seem, according to certain surveys, to take place in industry, expert systems have also been developed in the tertiary sector (banks, insurance companies and so forth). We wanted to work in environments with which we were familiar, not only to be able to choose our projects better, but also to be able to situate them in relation to broader or more long-standing tendencies affecting the industrial world. We had, moreover, already studied such tendencies, which have been the subject of a great deal of research during the past few years[13].

We shall now briefly present the experiments, other than Naval, which played a major part in our thinking. The following paragraphs outline the industrial contexts and the fate of the projects that are discussed in later chapters.

3.2.1 The TOTEM project: preparing production routing

The aim of the project was the realization of a system to prepare production routing in two workshops (rolling and wire drawing) in a precious metal processing firm. These routings were normally prepared by production planners, who provided the basic knowledge for TOTEM. The project was led by one of the company's engineers who had a broad perception of the problems involved in production management in the metallurgical industry. After three years the system was in daily use, but in a modified organizational context.

---

13 See Cohendet P., Hollard et al. (1988).

3.2.2 The GESPI project: assigning routes and tracks in a large railway station

This time the aim was to develop an expert system which would draw up the daily plan of routes and platform tracks for main line and commuter trains in a large railway station. The plans were produced by the station's "planning" department, which then transmitted them to the dispatchers. The project mobilized several actors over a period of four years. It is operational today, after undergoing several modifications. Besides the difficulties encountered in the modelling phase, the project is noteworthy for having revealed the peculiarities of the "planning" department.

3.2.3 The Cornélius project: maintenance of a flexible cell

This project dealt with a problem frequently encountered in AI applications: breakdown diagnosis, or maintenance support. The facility used for the experiment was a flexible cell in a machining workshop of a high-tech mechanical firm. Cornélius was never made operational, but was studied over a long period and raised several interesting questions. What problems does the transfer of maintenance expertise pose? What relation is there between these problems and the way in which the system's knowledge and reasoning were modelled? The product of a technical crisis, Cornélius did not manage to survive its partial resolution.

3.3 A methodology for monitoring projects

In our research we opted for the basic principle of studying projects underway. This implied periodic meetings, from the birth of the project through to its conclusion or discontinuation, with its designers, leaders, experts and users. It also implied a long-term commitment; in three cases the research lasted for over two years. We were particularly closely involved in Naval, where we participated in the design of the system and the management of the project. The experience gained in this way enabled us to focus more closely on the essentials of the other projects. In the case of Cornélius, however, we were forced to rely largely on retrospective accounts. Our general method can be summarized as being monographic and longitudinal[14], yet it seems equally important to note the *de facto* use of several research methods, from the close involvement of the researcher to a more general historical approach.

---

14   Pettigrew (1990).

With Naval our approach was that of participant observers, a method which we have often used[15] and which is characterized by the involvement of researchers in the transformations they are studying. The researchers are, however, individual actors with their own status and freedom, granted them by the other actors involved. One could say, very briefly, that whereas outside observers try to make themselves invisible so as not to influence the phenomenon they are observing, researcher-interveners have a different objective. Knowing that the process they are studying can take one of several possible courses, they need, in order to understand its progression, to be familiar with the alternatives, and sometimes even to have tried one of them. They are then able to highlight the different points of view or the forces which are effectively active in a process of change. This direct experience with Naval naturally conditioned our framework of analysis. First, we gained knowledge enabling us to be more attentive to certain details that may well have been of particular significance in the outcome of the project. We were able to rid ourselves, more easily than by mere observation, of the generally accepted distinctions between the technical and social aspects of the project, for example. It was thus possible to understand the interactions which were produced by each of its stages or which weighed on every choice that had to be made. But most importantly, our intervention strengthened our conviction on one point: *the necessity of allowing an adequate time-frame for analysing projects*, with "adequate" meaning the time required to perceive the most fundamental issues. Six months afterwards, our interpretation of the Naval case is totally different, and the reasons for it being discontinued can only be grasped if we piece together the jigsaw for an overall view of the different actors' involvement.

In the other projects we limited our involvement to that of observers, although studying a project underway and trying to understand events as they occur, without being actively involved, is no easy matter. Researchers have to be sufficiently familiar with the multiple aspects of an industrial context, and try to perceive the various options that are chosen. Moreover, they face the problem of being influenced by the views of their different interlocutors, although there again, time is a valuable ally and the dynamics of the project do reduce the chances of this happening. There is an obvious difference between meeting the different actors of a project *at a given moment of its development*, and seeing them regularly throughout its progression or decline. In the second case, the dialectic is more intense, as one has to constantly take into account the discrepancies which may exist between opinions and facts, between perceptions and actions. A difficulty which is under-estimated at one stage will not be ignored when it arises again later, and the researcher is able to perceive the evolution of the actors throughout the process with its changes and its crises.

---

15   Hatchuel A. (1988a).

3 Research material: expert systems in the industrial world   23

It was this type of unexpected problem which allowed us to test the hypotheses of expert systems, that is, the discourse of this approach on itself, outside of any context, and which served as a reference for the project leaders.

3.4 Three basic hypotheses: rationalization with a participatory nature

We have already referred to two specific principles of expert systems, which allowed us to introduce certain notions such as knowledge bases or inference engines. We can consider that there are in fact three basic hypotheses which served as a questionable, although acceptable, reference for the promoters of expert systems. Moreover, these three hypotheses play a greater symbolic role than do the computer mediums which are used for the realization of the systems and which have evolved and sometimes changed radically. It is, for example, significant that the concept put forward in the mid-1980s of computers dedicated to expert systems was abandoned without this challenging the approach itself. It is also noteworthy that the basic hypotheses highlight the attraction of the approach and the effort it demands in studying expertise.

3.4.1 Appealing hypotheses

The first two hypotheses correspond to the two principles already presented, while the third concerns the status of the expert.

Hypothesis 1 states that expert systems aim at "capturing" the natural knowledge of human experts, in the form closest to that expressed by the expert. It is then possible to "imitate" the reasoning of certain individuals to resolve certain problems by means of computers.

Hypothesis 2 associates expert systems with a particular form of data processing, or more specifically, programming. The general structure of the software separates "knowledge bases" from "reasoning". This division makes it possible to gather the first independently of the second and thereby to ensure that the knowledge base is autonomous and modifiable, without affecting the reasoning system.

Hypothesis 3, which we have not yet discussed, specifies that the expert who provides the knowledge must also validate the results of the software. This hypothesis is almost a corollary of the two preceding ones, but nevertheless merits separate treatment. It means that the "cognitician" (this word has practically disappeared today, although it was in current use in 1985), the expert system specialist, merely uses the know-how of recognized experts in a given subject domain, and that these experts alone are capable of judging the results obtained by the system.

The recognized advantages of expert systems are based on these three hypotheses. The separation between knowledge and reasoning should facilitate the rapid and progressive development of the system, and universal reasoning tools can be proposed (e.g. inference engines, "intelligent" classification). The user-friendliness of software designed in this way, the possibility of disseminating knowledge where it was formerly inaccessible and the evolutive nature of knowledge bases, are often claimed to be considerable advantages.

But the most appealing characteristics are perhaps to be found elsewhere, and *chiefly in the apparent ease with which these projects can be launched*. The feasibility of the project, a recurrent obstacle in industrial life, never seems to be a problem; since it is only a matter of harnessing available knowledge, there seems to be nothing to invent, nothing to find. Better still, these projects have modest ambitions: they need only perform as well as human experts. At the same time there is uncertainty about the experts' attitudes. Are they willing to divulge their secrets? Should the "cognitician's" main virtue not be diplomacy? One thing is certain, and that is that the experts are needed, not only for their consent but also for their lasting commitment and their active participation in the work of modelling and validation. *An ES project is thus part of a rationalization procedure which has to reconcile computer sophistication and a participative approach.* Its leaders are obliged to listen attentively and studiously to the experts explaining their know-how, and then to accept the same experts' verdict on the relevance of their system. Thus, there seems to be little chance of the project being diverted to technocratic ends.

3.4.2 Steps in a process of discovery

The projects studied will demonstrate that the advantages discussed above are not always accessible or, more surprisingly perhaps, that they are sometimes only attained if one or more of the basic hypotheses is ignored. This is due to the fact that *the nature of expertise mobilized in industrial firms, the dynamics in these firms and the impact of expertise on organizational or relational processes, are themselves largely ignored by hypotheses such as these*, whilst in fact they play a significant role in the outcome of ES projects.

We shall first show why the basic hypotheses of expert systems are based on a very restricted conception of knowledge, which will in turn enable us to distinguish three broad types of know-how used in industry: *"doing know-how", "understanding know-how"* and *"combining know-how"*. These three types of expertise correspond to ways of mobilizing very different types of knowledge, are not composed in the same way, display knowledge which is not articulated or heterogeneous to the same extent, and present very different forms of validation. We shall then see that "the imitation of expertise", because it is a process of automation of knowledge, is possible only at the expense of the "active" transformation of this knowledge; it is hence in itself a creator of expertise. Thus, the idea of the "cognitician" collecting expertise as it drips from the tree, is hardly applicable here, and the process of vulcanizing rubber is indeed a more suitable image than that of harvesting hevea! These dynamics are not without consequences on the network of actors involved in the project; with the restructuring of expertise, tensions quickly arise which influence its eventual success or failure. We are hence at the heart of the issues at play in a group when there is a question of creating or modifying the distribution of each actor's expertise, and hence the contents of shared expertise. In the projects we studies some of these problems were immediately perceived by the expert system specialists, most often without any clue as to their common origins or their articulation. In fact they derived from a poor understanding of the nature of knowledge and of expertise in organizations. Thus, over and above a study of expert systems, we were convinced that we could deal with a neglected aspect of organizational life, at a moment when this aspect was of greater strategic importance than ever before.

Chapter 2
Artisan, repairer, strategist
Different facets of expertise

Our first journey into the history of the four projects studied will focus on the contents of the expertise that was progressively analysed and modelled. It is important here to investigate the different networks of reasoning, action, knowledge and problems involved, and we shall do so by looking at each of the projects individually. *The relevance of such an approach lies in the fact that it is within the very core of expertise that the hypotheses of expert systems are put to the test. It is similarly at that level that it becomes progressively clearer who the expert actually is, and in what framework of action he or she intervenes. Finally, it is there too that the main tensions of the project are born.*

What we notice first is the variety of knowledge and expertise involved in the different problems, for it contrasts sharply with the limited nature of the basic hypotheses of expert systems and is, in particular, incompatible with the assumption of a separation between knowledge and reasoning. This type of conceptual distinction is indeed ambiguous: is "reasoning" not in itself a form of knowledge and, in certain cases, is it not the most important element in expertise? We shall see that this question can serve as a guideline for identifying the three main types of know-how which we call *"doing know-how"*, *"understanding know-how"* and *"combining know-how"*. Each of these types is distinguished by a specific type of organization and utilization of knowledge; and even if each of them corresponds to a particular type of situation, they are all present to varying degrees in any form of expertise. It is by understanding which part of each of these types of know-how is required for an activity, that one can define the nature of the technological constraints characterizing the activity.

1 Knowledge and reasoning: where is the boundary?

There is no reason for the notions of knowledge and reasoning not to be considered as equivalent *a priori*, but expert systems are based on the formal organization of this distinction. Reasoning is said to take place in the "inference engine", the device which makes deductions and conclusions and of which the almost universal nature is emphasized. Although this universality has been contested and theoretical debate on the matter abounds in the AI domain[1], one cannot deny that the software which made the development of expert systems possible, owes its existence to the distinction between knowledge and reasoning.

1.1 A practical but limiting distinction

AI theoreticians have shown particular interest in this distinction, for it allows them to explore new avenues in knowledge representation without changing the structure of expert systems. Hence, recent developments have dealt with the multiple ways of representing expressions of the "A = B" type[2], when these no longer signify "All As are Bs" and may, for example, mean that "certain As are sometimes Bs". This kind of formalism may make it possible to constitute more flexible fact or knowledge bases which accept less rigid statements than ones like "the cholesterol level is greater than x" or "at least one of the resistors in the circuit is burnt out". Such developments can be made without any major modifications to the inference engine that has to handle new knowledge.

Even if the distinction between knowledge and reasoning makes sense with respect to software, it is somewhat deceptive. The idea of "raw" knowledge is meaningless, for as soon as it is formulated, knowledge is expressed in a form structured by implicit rules for the creation of statements, or by principles of causality. Most expert systems are designed to function only with knowledge written in the form of rules of the type "if A = B then C = D". If one looks at apparently unstructured knowledge such as a list of different types of cats, for example, one sees that even at this level of simplicity it involves the implicit rule of reasoning "if A = a cat, and if B = a cat, then (A,B) = a group of cats". *Thus, knowledge is always constructed with the implicit use of rules.*

The degree of basic interaction between reasoning and knowledge varies from one case to another. We shall see that certain expertise can be characterized by this interdependence, which increases in proportion with the complexity of the expertise.

---

1 Sombé L. (1988). Léa Sombé is the pseudonym of a group of AI researchers.
2 Sombé L. (1988).

1.2 Classification of expert systems: getting back to the nature of expertise

During the past few years, an abundant literature has provided varied descriptions of expert systems. These texts present a broad range of applications, usually sorted into broad subject fields for implementation (industry, banks, insurance, etc.). Other authors have tried to define a more fundamental typology based on the nature of the problems dealt with. They generally result in classifications distinguishing, for example, diagnosis support, scheduling support, procedure control support, design support, decision support, and so forth.

This type of classification has the advantage of making a lot of sense for potential users, but it also has the drawback of saying nothing about the nature of expertise dealt with, and notably about the way in which knowledge and reasoning are represented in each case. Moreover, a classification by type of problem masks an essential fact: *any particular problem, for example one of scheduling or diagnosis, may be solved in different ways, depending on available expertise.*

An AI specialist knows that he may well encounter very different types of difficulty from one problem to another, difficulties which may challenge the very idea of expert systems in the traditional sense of the word. However, these problems are expressed in the form of debates on formalism or computing languages, both of which remain obscure to the layman. Thus, at the start of the 1980s a distinction was made between two methods of representing knowledge. The first, representation by "production rules", corresponded to the accumulation of statements of the "if A then B" type. The second used "object oriented" languages, and implied the introduction of "objects" (e.g. a machine, a contract, a class of individuals) characterizable by a set of attributes or rules for operation.

Naturally, these issues remain abstruse to the novice until he or she is informed which expertise is best described by a certain approach. Yet behind such technical debates lay a new awareness, amongst specialists, of the existence of a large variety of expertise. Rather than a classification in terms of problems or fields of application, it was a classification by types of expertise that revealed the peculiar nature of the initial hypotheses of expert systems. *This type of classification is, however, basically a theory of relations between forms of expertise and courses of action.*

Based on the cases studied, the latter type of classification did seem to us to be possible, using three symbolic characters – *the artisan, the repairer, the strategist* – each of which suggests a type of relationship between knowledge and action. The classification can be justified by the fact that it concerns three principal figures in industrial history, and that it is related to certain philosophical traditions. However, any further discussion on these particular subjects is beyond the scope of this book. It is rather by delving to the core of the experiments studied that we will be able to define the most specific expertise of these three characters.

Our analysis will start with the TOTEM project for preparing production routings. Here we see how an expert's know-how can be described as "doing know-how", consisting mainly of an accumulation of known and memorized procedures such as those which constitute techniques or crafts.

The Cornélius project for diagnosing machine failure introduces us to the second type of expertise, that of the repairer. The expertise in this case does not consist of activating a sum of procedures, but of solving a riddle for which there is always a solution. We call it "understanding know-how".

Finally, with Naval, a project for scheduling the use of oil rigs, we enter into the changeable world of the strategist, where the choice of tactics, compromises and priorities is paramount. "Combining know-how" is required to find an answer adapted to each situation, without any guarantee that such an answer exists.

2 "Doing know-how" or the artisan's expertise

We shall use the TOTEM project[3] as a basis for analysing the first category of expertise which we have called "doing know-how" and have symbolically attributed to artisans.

2.1 The preparation of routings for metal processing

The aim of the TOTEM experiment was the automatic generation of production routings for an industrial unit consisting of a rolling and a wire drawing workshop processing precious metals, notably silver, gold and platinum.

A production routing is a set of information specifying the raw materials and different operations required in a particular production process. The information is contained in a document issued by the production planning office to the relevant workshops, which then have to follow the prescribed instructions and sequences. In a workshop which mass-produces automatically a small range of items, the routing is almost always the same and is implicitly incorporated in the organization of the machines themselves. However, in the type of industry in which the TOTEM project took place routings play a significant role.

The factory concerned produces to order, and its industrial resources permit the manufacture of a wide variety of products. It can, for example, produce strips, sheets or wire, in a variety of alloys, shapes, sizes and metallurgical states. Conse-

---

3 See in Part Two, "Totem – The reconstruction of production planners' expertise".

quently, the range of routing possibilities is vast, and for each new order a specific new document sets out the required production process.

Even though the notion of "routing" corresponds to a long-standing practice in the metallurgical industry, and the presence of a planning office for preparing routings is the most common organizational solution, the details provided on these documents vary from one situation to another. *These variations depend on the production planners' expertise, or more precisely, on the distribution of expertise between the planning office and the workshop.* Thus, the contents of a routing represent the difference between what the workshop can do on its own initiative, and what it has to do according to instructions on the document. The know-how used in the preparation of a routing does not therefore correspond exactly to all the know-how required to carry out the operations described; the routing is not an "operation manual" describing in detail the tasks to be executed by the workers on the shop floor. We shall look at two examples encountered during the TOTEM project.

In the case of a rolling workshop, the choice of a piece of metal to be used as a base implies several alternatives. For example, to manufacture a 0,5 mm thick sheet, it is possible to start with a metal of any thickness greater than 0,5 mm. The same applies to the width and length. Production routings can therefore specify the dimensions of the base or leave this type of choice up to the workshop. Similarly, production operations may be specified without reference to the machines to be used and, when a rolling operation can be performed on any one of several mills, the final choice will be left to the operators.

As the above example demonstrates, the term "doing know-how" does not necessarily mean a body of knowledge that includes the finer details of shop practice or of each act to be performed (this is moreover merely a theoretical limit, for even the instructions for assembling a piece of furniture in kit presume that the customer knows how to use a screw-driver). It must be seen to represent *a type of expertise which expresses, whatever the level of detail considered, the way in which certain transformations are obtained by familiar actions.* In other words, "doing know-how" allows one to determine the intermediary steps between the initial state and the desired state. One of the distinctive features of this type of expertise is that it can be described as a *set of situations* together with the actions which make it possible to pass from one situation to another. It is knowledge which is, understandably, easy to archive and effectively archived. Most industrial firms have libraries of routings, or reusable standard routings composed of the "question-answer" combinations which were found to produce satisfactory results in the past. A guide to home decorating or entertaining, or a famous chef's recipe book, are common forms of memorizing and displaying "doing know-how". This is the world of recipes, the archetype of what is called "technical know-how" in everyday language. It borders on the activity referred to by the ancient Greeks as "poiesis", that of someone who knows how to produce objects and in order to do so uses all the secrets of nature known to him or her[4].

2.2 The representation of "doing know-how": a typical structure for expert systems

Described in this way, *"doing know-how" is clearly suited to the basic hypotheses of expert systems*; it makes it possible – as in the case of TOTEM – to adhere to their main principles. Work with the production planners, who were naturally the routing "experts", was at the core of the system's development, so that it was effectively their practices which were reproduced at first. This was possible for a large number of products, precisely because the planners' expertise could, *in such cases and in such cases only*, be described by a structure of knowledge presenting three main features: the effect of learning, the effect of logical simplicity, and the effect of exploratory neutrality.

a) *The effect of learning*: Expertise is acquired mainly by means of the accumulation of problem-solution combinations, relating to a very large number of situations. The first obstacle encountered is that of "recording", peculiar to any approach aimed at the most complete accumulation of knowledge possible on a particular subject.

b) *The effect of logical simplicity*: For expertise to be described as "doing know-how", in our sense of the term, it is necessary to have progressively defined autonomous operational sequences, which then need only be linked or connected in order to obtain the desired result, as in the case of production routing. This form of expertise, which requires simple reasoning only, is an industrial and pragmatic ideal which is always aimed for in one way or another. It consists of selecting only that which can be expressed in forms that facilitate the connection and resolution of the questions to be dealt with.

However, in order to achieve this during the TOTEM experiment, *it was necessary to refuse the expert system some of the leeway generally given to human experts* and, for example, to define thicknesses as compulsory steps in the preparation of a rolling route sheet. Such rigidity allows the system to reason simply, and to determine the state of the metal to be used as a base, by working back step by step from the desired state of the final product. At each step the problem to solve will be the same: "given that the thickness of the final product is X mm, what is the thickness greater than X which makes it possible to obtain the desired result after only one pass on the rolling mill?". With the preliminary definition of inter-

---

4 On these different forms of activity and practical knowledge defined by the ancient Greeks cf. Vernant J.P. (1983).

mediary thicknesses one can therefore constitute autonomous sequences, where the ES is told that to obtain an X mm sheet in one step, it must start with a Y mm thick sheet.

Structured in this way, the expertise in a routing could be expressed in the form of simple knowledge. Had this not been possible, the determination of intermediary rolling levels would have presented a new problem with every order, to be solved by means of more elaborate methods. These would then have had to be included in the expertise required for the routing. Such an extension would have spoiled the logical simplicity of the know-how and would have led TOTEM towards another type of expertise, that of "combining know-how", which we examine later.

c) *The effect of exploratory neutrality*: What is the quality of a routing obtained by a system like TOTEM? Is it the best, the simplest, or the most economical? By the very structure of "doing know-how", an expert system of this type may well function without there being answers to questions such as these. It may simply be said of the routing prepared by the system, that it conforms to the operational sequence realized by means of the know-how provided[5]. The know-how used by TOTEM was no doubt developed with one of the above-mentioned objectives in mind, and we know that production planners can have several objectives when preparing a routing, such as the choice of the most efficient machine or procedure, or the optimum use for available raw materials. But once it has been acquired, the expertise may show no apparent sign of these goals, and the generation of a routing will be obtained without any reference to the objectives that initially served to develop it. This is what constitutes what can be called the effect of exploratory neutrality, since knowledge is only utilizable by means of rules of formal logic. It is also another way of showing that the independence between knowledge and reasoning is acceptable here.

Learning, logical simplicity and exploratory neutrality are the basic conditions for expertise to be represented by means of the main characteristic tools of expert systems, i.e. a knowledge base constituted by statements of the "if A then B" type, a fact base composed of propositions of the "A = B" type and an inference engine which uses rules of ordinary logic. *But, although the generation of routings fits fairly easily into this type of framework, the TOTEM project demonstrates the first limits of structures such as these.*

---

5   This routine aspect of know-how led the ancient Greeks to doubt the fact that it was effectively expertise (Vernant 1983), since it was possible to learn or to use it without understanding it.

2.3 The role of calculations and technological models

Several problems arise when the representation of "doing know-how" demands the use of reasoning which is not logical, but numerical, and when the realization of a relevant system requires the development of a technology where nothing but a collection of techniques exists.

2.3.1 From empirical expertise to the technological model

We shall start with the second problem, certainly the trickiest, which can be approached by introducing the following question: *"What distinguishes an expert system which prepares routings, from a mere routing library?".*

The answer is that an expert system has no set definition of a production routing, nor a list of routings from which to choose. Each case is a new problem for the system to solve, and nothing prevents it from finding a solution which has already served in the past. It builds routings by means of rules stimulated each time by the same goal, i.e. obtaining a product in conditions prescribed by the customer's order. But these rules, and that is the main point, do not just form a grid which automatically links a sequence of operations with a particular type of order. They must, on the contrary, organize the progressive construction of the routing by means of diverse knowledge which may or may not be applicable to the relevant problem. Take for example the following proposition: *"after rolling the metal is in a hard drawn state"*. This is a standard notion in metallurgy, which expresses general knowledge of the internal structure of metals, is independent of the characteristics of the product to be produced, and therefore may well be mobilized at very different moments of the reasoning. In contrast, propositions may be required which will only be utilizable at a specific stage of the processing, since they depend on the nature of the operation carried out (e.g. propositions relative to the geometric characteristics of the metal). Thus, the representation of certain expertise demands the implementation of several levels of rules, from abstract or conceptual rules structuring the main stages of reasoning, to the most specific rules that are only used to solve localized problems.

It is in the development of general rules structuring the expert system's approach, that so-called "added intelligence" of the system as compared to the experts' know-how, is constituted. Relations of this kind are not necessarily perceived or expressed by the "experts" concerned, and some of the abstract concepts involved are developed during the project and not forged naturally by the practitioners. The relations and the concepts on which they are based constitute, in the strict sense of the word, *a technology*, as opposed to the particular techniques which a technology encompasses. In this way "doing know-how" which, historically, has been associated with engineers and contrasts with the more contextual expertise of operators, is formed. In TOTEM the preparation of routings for the rolling workshop produced several conceptual creations of this type. It was necessary to define distinct areas of expertise corresponding to large families of routings, to homogenize certain apparently different notions, or to restructure the main technical stages of production. Thus, with the example of TOTEM, one has to start qualifying the idea that an expert system is a mere imitation of experts' knowledge, although the situation does on the whole lend itself to this type of hypothesis.

2.3.2 The role of calculations: at the limit of "doing know-how"

Amongst the conceptual relations involved in the formalization of know-how, those which are expressed in a quantitative form play a particular role. They express the necessity of using knowledge or reasoning that does not depend exclusively on the logic of propositions, unless – and that would be absurd – it wanted to rediscover the properties of algebra or geometry. For example, in the preparation of certain routings, it was necessary to choose the most suitable metal as a base, but the experts' explanation of the rules governing such a choice were inconclusive. Certain practices had established themselves without very firm grounds, and different answers were supplied by various experts. Faced with the need to systemize the modalities of these choices in TOTEM, the project leader developed a series of geometric calculations in response to the question asked. He thereby introduced the gap which exists between expertise constituted by an articulated series of practical solutions, and the reformulation of this expertise in more general terms, which allows for a unified and systematic approach to the problem.

The experience with the generation of routings is therefore a perfect illustration both of the concept of "doing know-how" and of its limits. It consists of systemizing expertise which produces operational modes where the result of each step is controllable, and which can also be general enough to be applied to new products or new routings. This type of knowledge fits easily into the conceptual and computer mould adopted for expert systems. Nevertheless, the slow progress of the TOTEM project and the uncertainties which reigned over the final configuration of its insertion were due to the fact that, even in the field of expertise where it belongs, the traditional philosophy of expert systems left several questions unanswered. We have just seen why the modelling of know-how is not merely a question of logic and how it also needs to resort to conceptualization and abstraction. It may then be a matter of reconstructing, from fragmented empirical know-how, the elements of coherence and systemization of a technology. Such ordering, sometimes inseparable from computation, is not always easy, or possible, and extends well beyond the initial objective of gathering expertise.

The implications of these processes can now be perceived. Did the TOTEM project experience introduce into this firm a broader and more systematic vision of the preparation of work? Did TOTEM's designers lay the foundations for a new type of engineering? If so, who were the actors in this mutation and what was the long-term industrial significance? We shall look at these questions later on in this study. First, we need to investigate the expertise which differs from the stratified ordering of "doing know-how".

3 "Understanding know-how" or the repairer's expertise

We have seen how the artisan's expertise is expressed directly and simply. This type of knowledge states how to go from one object to another, which path to take, and which resources and rules of thumb to use; it is clear where the process must start and when it is finished. The artisan's experience can be ordered into rules, and a variety of situations dealt with by memorizing solutions. *In contrast, the repairer's expertise[6] is more complex; it cannot be laid out in a straight line, and in each new situation it intermingles action and investigation in an ever-changing pattern.* The repairer examining a machine and the cook starting on a difficult recipe both have cause for some anxiety, but the former cannot forget that it is not enough to be careful and attentive, for there are failures which are incomprehensible and reparations which are unsuccessful.

3.1 "Doing know-how" and "understanding know-how"

The repairer has to restore to its original state, or as close to it as possible, an order which has been upset, diverted or deformed. The difficulty seems to lie in the fact that before a problem can be solved, its origin has to be found. This analysis is correct, but nevertheless under-estimates the subtlety of the problems encountered, for the repairer has to combine several concurrent or complementary procedures.

a) Even if searching for the cause of a phenomenon requires a capacity for induction (certain signs evoke certain causes), a strategy of investigation is also required. When known signs of failure are insufficient to reach a conclusion, it becomes necessary to find new ones. But what signs, and in what order?

b) Determining the exact cause of a phenomenon is desireable, but not indispensable, for a repairer. Sometimes he needs no more than brief indications in or-

---

6 We have borrowed the image of the repairer from Isaac Joseph (Joseph I. 1989) and note that the technical, civil and contractual skills that he attributes to a repairer can all be involved in the reconstitution of the latter's expertise.

der to make a decision. For example, to repair a mechanical unit it is possible to change a sub-unit, without worrying about the elements of this sub-unit causing the problem.

c) Finally, it is possible to act without any certainty that the part is faulty, since substituting the part is likely to produce two types of result: that of repairing the dysfunction and that of simultaneously proving that the replaced part was or was not the source of the problem.

From their infancy, expert systems found in the expertise of medical practitioners a perfect field for experimentation, and the Mycin project for diagnostic support[7] is one of its better known pioneers. In the world of industrial installations, the repairer *par excellence* is the maintenance fitter, the "doctor" of machines or procedures. The idea spread rapidly that breakdown diagnosis, which is maintenance fitters' basic expertise, was highly suitable for the development of expert systems. But are the general concepts on which expert systems are founded, and of which we have just measured the suitability and limits for "doing know-how", as well suited to the "understanding know-how" of a physician? The case histories mentioned above seemed to demonstrate that they are, provided, however, that medical expertise and industrial expertise resemble each other closely.

We shall see that such similarity is only obtained at the price of a particular formulation of the experts' knowledge. This has not only led computer specialists to create languages dedicated to the problem, it also implies that the context of the repairer's analysis and action can be defined in practice and with a large degree of precision. *The repairer's expertise cannot be defined out of context, it requires that the practical and social framework of the reparation be constructed simultaneously.* That is what we shall see with the Cornélius project[8] for failure diagnosis in a flexible cell.

3.2 Cornélius: maintaining a flexible cell

Although automation has allowed for flexibility in many recent industrial systems, is also involves a number of drawbacks. The complexity of facilities is often accompanied by risks of reduced reliability which decrease utilization rates, that is, the actual availability of the unit. There are several remedies to contain or lessen this risk, one of them being, of course, the reparation of the facility as fast and as efficiently as possible. However, this implies determining rapidly and accurately the steps that need to be taken. That was precisely one of the objectives of

---

7   Shortliffe E.H. (1976).
8   See in Part Two the chapter "Cornélius – Fragmented expertise of maintenance specialists".

the project undertaken in 1987, on a machining flexible cell in a large advanced mechanics firm that produced parts for aircraft equipment. The cell functioned in three eight-hour shifts and was sufficiently automated not to require an operator at night. Since the factory's entire output had to pass through this point, the cell was in effect a strategic bottleneck, where any gain in reliability would directly affect overall productive capacity. The Cornélius project therefore aimed at the creation of a diagnostic support ES that had to be armed with available knowledge on the cell. Three types of actor could each possess a different part of this knowledge:

- the highly-skilled technician from the central maintenance service who had installed the cell and had had overall responsibility for it for three years;
- the maintenance fitters from the workshop in which the cell was situated, who were responsible for its maintenance;
- the operators of the cell who were authorized to restart it in certain cases and to carry out quality control on workpieces during production.

It was with the first of these actors that the project was launched and that the first "knowledge bases" were built. From the outset the choice of the "expert" seemed to be a key question in the project. This was not the case with the TOTEM project where the production routing was an object that was well situated organizationally and where it was possible to identify expertise and the experts simultaneously. The "object" maintenance did not have the same type of identity and the choice of an expert was in fact a way of defining it. This choice was by no means insignificant and the project was to come up against one of its consequences. It became clear after some time that the diagnostic procedure adopted by someone who was mainly interested in understanding the internal mechanisms of the facility, was more analytical than that of an operator used to reacting to breakdowns and sorting out problems as quickly as possible, rather than explaining them.

A discussion on maintenance know-how will enable us to better understand this distinction.

3.3 Appropriate knowledge for diagnosis: the development of a suitable physiology

A priori there is no reason not to extract knowledge in the form of "doing know-how" from a maintenance expert. However, this idea, although theoretically acceptable, seems barely practical when the aim is modelling a diagnostic procedure. Inferring the causes of a problem from signs which can be observed, implies a description of the normal functioning of the unit, that is, of the strings of causality of the system under consideration. This step is an essential condition for the separation of possible causes and effective causes, leading to a diagnosis.

3.3.1 Possible causes and effective causes

Take for example a typical problem encountered by a motorist[9] whose car will not start. The only sign of failure is that "turning the key has no effect". With this information, the driver can usually deduce that the breakdown has one of three sources: battery failure, starter failure, or failure of the electric circuits linking these parts. But this list of causes does not include all the knowledge of an expert repairer. He also knows that apparent failure of the starter may be the mechanical result of one of the other causes; in other words that the three components are functionally linked. The causes mentioned initially are therefore only "possible" causes, and in order to identify "effective" causes, it is necessary to take into account the theoretical relations between the parts that are out of order. Thus, with diagnosis, the knowledge that is described at first is clearly not "diagnostic" expertise as such, but the theoretical functioning of the unit to be analysed. One could even say, by pushing this logic to its extreme, that perfect knowledge of the theoretical functioning would be the only knowledge required, with the rest being reduced to systematic exploration of causes. Significantly, this analysis has been taken up on a technical level in the computer domain.

Classical expert systems, which correspond to the model given in the last chapter, soon proved to be difficult to manipulate when it came to failure diagnosis. As we have seen, the reason was that at least two types of rule had to be represented: rules describing the theoretical functioning of equipment, and rules which questioned this functioning. However, the formalism of logical inference required for the second type of rule rapidly becomes repetitive and redundant with descriptions of a core of stable interactions. Consequently, programs capable of handling this second category of knowledge more easily had to be developed, leading to a de facto breach in the supposed universality of expert systems as presented in the software proposed to firms at the start of the 1980s. The formalism of a proposition representing isolated elements of expertise, had to be replaced by new coherence built around "objects".

3.3.2 Object and diagnosis oriented languages: articulating knowledge

This category of software is generally qualified as being "object oriented", which simply means that it facilitates the manipulation of relations between "objects" (with the word taken in its broadest sense, a rule being an object). But it is less the programming possibilities offered by the software than the general approach underlying it which interests us here. The latter highlights the need to take into account knowledge which is not characterized, like "doing know-how", by learning and mere logic, and which leaves more room for the description of a mode

---

9   Laurent P. (1987).

of functioning or a dynamic structure. It is therefore not surprising that the first languages of this type were simulation languages. Their main objective was, for example, to simulate the process of operations in a workshop during a given period. These languages provided the advantage of having modelled typical objects such as machine-objects, parts-objects or queueing-objects in a preliminary phase, and so of representing the habitual mechanisms of production in progress. Thus, when diagnostic support demanded the description of the main aspects of a technical process, the idea of tools specifically adapted to this end naturally took shape. It was even thought that diagnosis called for the modification of the basic hypotheses of expert systems, seen as being far too universal. The programme Cornélius, selected for this project, is part of the software dedicated to diagnosis.

3.3.3 Options in modelling: sources of expertise

Representing an operation so that it can be examined and explored, appears clearly as an objective of diagnosis. But it has to be made operational, in other words given limits and criteria of relevance. To what point must an operation be described? Must all the components that could possibly fail, be described? Must one describe only those whose efficiency can be tested, or rather those which one can replace oneself? How can the facility be divided into relevant parts? These questions of limits are compounded by other more complex ones. Could one easily construct a logical interrogation of the causes of failure, since the concern for detail in the diagnosis can be faced with the impossibility of qualifying the state of certain components? It was questions such as these which lay at the heart of the project. The answers provided were drawn from various sources: besides the maintenance expert mentioned above, the actual context of the flexible cell and several preliminary studies helped to determine options for modelling.

a) The unit was already equipped with a control system enabling the operator to identify dysfunctions. With this information it was not possible to detect the reasons for a failure, but the system did offer a useful sub-division of the unit. Due to the complexity of the facility, only the following three functions, recognized as the main sources of problems, were studied in the project: palletizing, which provided the supply of workpieces and the unloading of finished parts; the tool magazine containing the different instruments used during very varied machining operations; the automatic control of the speed and positions of these instruments.

b) Before the start of the project, a maintenance manual had been written by a work group comprising a workshop supervisor, two operators, a maintenance fitter and a representative of the central service for assisting with the integration of new facilities. This preliminary work had already produced a description of the components and procedures of available tests; it had made it possible to clear the ambiguities that resulted from diverse definitions formulated by the different actors. The situation therefore lent itself to the formalization of a more advanced stage, that of the writing of dynamic interrogation rules. Several obstacles were, however, encountered.

3.3.4 The discrimination of efficient causes: who needs help?

Once the facility and relations between its main elements had been described, it was necessary to define rules to discriminate between normal and deviating behaviour of each component. These rules were to serve as a guide for interrogation and for the selection of tests to be carried out. The Cornélius project experience shows that the constitution of these rules was not easy and that it was often necessary to backtrack. Two objectives guided these attempts: *reducing the number of questions put to users by the system before it reached a conclusion, and ensuring that these questions were asked in a comprehensible order, adapted to the intended approach.* For example, when the system knew that several components could well be the cause of a failure, but was unable to prove which were in working order, it proposed a series of questions so that the user could test each of the components. The order of the questions had to take into account several criteria. For example, it was better to first ask questions which could easily be answered and which were "informative".

The "expertise" required here clearly related to two types of problem. The first consisted of structuring the questions according to the user's ability to perform tests, so that the relevant maintenance fitter had to be considered in economic and organizational terms. The second consisted of analysing the sequence of relations leading from possible causes to effective causes. The study of components with feedback loops was particularly problematical here. These components were equipped with their own control system, and this feedback loop acted as a "local diagnostic device" which in turn was also subject to failure. The modelling of the sequence of causes between a component and its automatic control necessitated the possibility of escaping from the circularity of the phenomenon, notably by tests where each one could be studied separately. However, such complex tests seemed way above users without any specific training in this field. At the same time, the project highlighted the advantages of new design principles which were to permit the introduction of simple automatic control tests in future facilities.

Thus, implementing the Cornélius project proved to be far more complex than anticipated, and the effective realization of an expert system did not go strictly according to plan. In fact, most diagnostic support systems like Cornélius are designed to be used as a means for transferring the experts' knowledge to actors who do not have such expertise. *However, the word transfer is misleading here, for it is not merely a matter of a shift but of a complete reconstruction of available expertise, according to models applicable to the actors that are to receive it.* This notion, which in principle applies to any transfer of expertise, is particularly relevant here due to the specific structure of "understanding know-how".

3.3.5 Diagnostic support systems: understanding in practice

Although they showed interest in the Cornélius project, the machining centre operators concerned often remarked that they were used to intervening very rapidly. They feared that the system, while providing them with additional expertise, may nevertheless be rigid and slow. This argument is not to be taken solely as one of technical acceptance, in other words, the result of misgivings about the system's speed. It is rather by looking at the connection between diagnosis and corrective maintenance that the operators' apprehensions can be understood. In passing, it seems relevant to mention that we encountered the same problem in another maintenance support project, with lift servicemen. In that case the system designers also discovered very quickly that the servicemen's approach could not be described in diagnostic terms only.

When faced with an urgent problem, operators do not always try to validate their idea of the cause of the problem; they rely far more on past experience. If the situation reminds them of the cause of problems encountered in the past, they first proceed as though the cause were the same, and then continue their diagnosis if this proves incorrect. A similar approach consists of the systematic examination of certain simple components, because the solution is so quick that "it would be stupid to miss it". Thus, it is clear that the introduction of corrective maintenance reasoning requires a broader conception of the expert system's relevance. New requirements are added, that the system has to take into account when putting questions to the user. *A diagnostic support ES could therefore very well start, not with a list of questions designed to define a cause, but with a list of acts to accomplish,* based on the preliminary analysis of the causes of the most frequent failures. As some authors have noted[10], the philosophy of this type of approach is not one of logical investigation of the causes of failure; in essence it is similar to the so-called Bayesian approach to decision making[11], in which information to be found (diagnostic refinement) and acts to be performed are formalized in the same way. In this formalism, the two modes of action are evaluated simultaneously by means of the actors' beliefs about the causes, and their assessment of the consequences, of their actions. At least this conception has the merit of making it easier to understand the difference between operators who often perform reparations by a quasi-systematic act, and maintenance fitters who only intervene on complex cases where determining the origin of the problem presents a major difficulty. This is familiar ground for anybody who regularly uses a GP and has often been told by his doctor "If you're not better in a week's time, come back and we'll do more tests". *A purely diagnostic approach would have proposed quite the opposite!*

---

10 Sombé L. (1988).
11 Raïffa H. (1970); Bourdaire J.M., Charreton M. (1989).

Our examination of the repairer's expertise progressively enabled us to take a more objective standpoint vis-à-vis conventional ways of considering the acquisition of knowledge in expert systems. Knowledge was no longer raw matter that needed only to be memorized; it appeared rather as the product of a theoretical construction where choices and raw facts were interlinked, and represented a distribution of tasks and issues peculiar to an organizational situation. Experts do not simply have to be questioned, their expertise has to be replaced in its own particular context. Everything they know is not necessarily interesting, and their way of seeing things may not be suitable if other actors are to use their expertise. Whoever they may be, users turn out to be deciders rather than consumers, avid for expertise as long as it is useful. *It is now easier to understand how maintenance expertise can be divided up between several actors, to the point of making the exchange of knowledge problematical. There are as many forms of maintenance expertise as there are ways of representing an operation and defining its dysfunctions.*

By looking first at "doing know-how" and then at "understanding know-how" we have seen how the expert's knowledge becomes more complex, and how the conditions of this knowledge intervene in the second type of expertise. At least there has never been any doubt on one point: *there was effectively expertise to record*. This hypothesis will however pose a problem when we look at the third type of knowledge, the type we have called "combining know-how", where expert systems are clearly meant to be strategists.

4 "Combining know-how" or the strategist's expertise

Our usual idea of a strategy is similar to that of a plan, although it is not exactly the plan of the artisan who prepares a production routing, like in the TOTEM project example. When organizing their course of action, artisans are guided by experience. *In contrast, strategists build a new story every time from the variable elements at their disposal, which they try to arrange as effectively as possible.* They need two types of expertise. The first is easy to define, and relates to the objects which give substance to the plan underway, i.e. tasks to perform, resources to mobilize, goals to achieve. The second type of expertise is more difficult to formalize and the following types of question arise: What is the art of combining, arranging and coordinating? To what types of knowledge does it refer? Can one talk of expertise in this case, or more simply of known and classified methods? What then is an expert system that generates action plans? Who is a strategist? A corporate leader, of course, but also a strategic planner or the co-ordinator of a team of professionals.

To shed light on these first questions, we shall examine two projects which demanded a great deal of time and effort. The first, Naval, dealt with problems of coordination and planning at a senior management level in a large oil company. The second, GESPI, dealt with the difficult problem of assigning platform tracks to trains in a large railway station.

4.1 Naval: planning the use of oilrigs

The birth of the Naval project[12] was closely linked to the nature of off-shore oil exploration and to the oilrig market. Rigs are essential in exploration and this heavy and complex equipment is generally owned by specialized firms which contract them out, for variable periods, to the oil companies that utilize them. The oil companies draw up annual exploration plans specifying the wells to be drilled on their different prospects. These wells are located at different points of a given geographic area, in which a number of rigs are supplied by the contractors. The gigantic size of most of this equipment leads to the creation of markets in fairly limited zones, for it is practically impossible to transfer an oilrig quickly and cheaply from one area of activity to another. An oil company's exploration programme will be feasible only if the company can find technically suitable rigs during the required period. These time constraints may be due to climatic, technical, or even contractual conditions if time commitments are made vis-à-vis host States or partners.

Most large oil companies operate in foreign countries through autonomous subsidiaries that are subject to local legislation and are often part of a consortium. Several subsidiaries in a single group may operate in the same geographic area and will therefore all use the same oilrig contractors operating in that zone. This multiplicity of needs and commitments is necessarily a generator of interaction, synergy, or conflict. In the company in which the Naval project took place, such interaction gradually resulted in the establishment of centralization procedures designed to reconcile the different subsidiaries' decisions concerning oilrig contracts. These procedures were progressively strengthened when the oilrig market was destabilized by several crises, notably in 1974, 1979 and 1982.

4.1.1 Crises on the oilrig market

The different oil shocks were evidently at the origin of these crises which, in turn, were directly due to the discrepancy between the time required to build an oilrig and the speed at which oil prices and companies' exploration programmes changed. Two particularly critical moments marked the 1970s and 80s.

---

12   See in Part Two the chapter "Naval – Undefinable expertise of strategic planners".

The first was caused by a reversal of oilrig market trends after a period of sustained growth. It started with the apparently consistent upsurge of oil prices which led companies to consider that an increasing number of prospects would be profitable, and therefore to allocate exploration budgets to them. The result was an inflated demand for oilrigs, and even tension on the market. Unable to respond immediately to this growth, the oilrig contractors tended, during crises like these, to increase their rates and to undertake the construction of new rigs. Some oil companies committed themselves even before the completion of the rigs, by signing long-term rental contracts at high rates. When, however, the price of oil stabilized, they had to reduce their exploration programmes and some of them found themselves bound by contract to expensive rigs for which they no longer had any use.

The second situation was in a sense a result of the first, since it was produced when the price of crude oil crashed after a long period of stagnation. Another type of crisis occured, for the companies had progressively freed themselves from their long-term committments and no longer hesitated to sign short-term contracts, at reduced rates. Decisions to explore were taken as seldom and as late as possible. The oilrig contractors' market shrank, and they were forced to reduce their fleets to a minimum, either by scrapping rigs sooner than planned or by withdrawing them from the market.

Although the latter type of crisis threatened the overall economy of the sector and ruined many oilrig contractors, it posed fewer problems of reconciling the decisions taken by each of the subsidiaries. The opposite was, however, true for the first type of crisis, where the co-ordination effort made full sense. Certain subsidiaries found themselves with unutilized and expensive rigs, whilst others needed rigs but could not find any. The parent companies consequently avoided any long-term committments and asked their subsidiaries to use, wherever possible, rigs rented by their counterparts in the same group. There were apprehensions about the future, and economic evaluations of prospects by oil companies often varied. The parent companies' concern for co-ordination was often resented by their subsidiaries, who wanted to act independently. They were not always able to explain to their local partners why they had chosen a particularly expensive rig (it was under contract to another subsidiary but was unutilized) whilst the market was depressed. Moreover, rigs already rented by the group were sometimes technically unsuitable (e.g. absence of certain functions) or unavailable at the most convenient times, and the subsidiaries used such arguments to support their decisions to rent new rigs. Programming the use of rigs on a corporate level thus implied compromise between contradictory or diverse positions, as well as a tricky process of negotiation. This was the context in which, in the mid-1980s, the idea was born of developing an expert system to produce these programmes, with the most coherent compromise possible between each subsidiary's constraints.

4.1.2 Planning drilling operations: between "doing know-how" and the search for a compromise

It may at first seem that the preparation of a drilling programme requires what we have called "doing know-how". An oilrig must be capable of drilling the well for which it is contracted, and must also be compatible with all the technical and security standards of the company. However, although these standards can preclude the use of certain rigs on certain prospects, they cannot guide the selection of one out of several suitable rigs. Had this been the case, the preparation of a drilling programme would have amounted to the mere application of technical knowledge. It was clear from the outset that such an approach would be impossible in the Naval system.

The existence of several acceptable candidates for a particular well was thus the focal point in the experts' and AI specialists' approach, which consisted of distinguishing two problems. The first was that of introducing an order of preference into the list of candidates, by means of certain criteria; the second used this scale to build a programme with the most "good qualities". In other words, if a programme respected all the rules defining "good" qualities, that is, all the company's wishes, it would be the best programme possible. Unfortunately, there was very little chance of such a programme existing, given the contradictory nature of most of the company's wishes. In practice it was only possible to look for a programme that incorporated the best possible compromise. An example of such a compromise would be one which transgressed the fewest rules, or the rules with the lowest priority, if such a scale of priority could be established.

The method followed by Naval therefore called for the use of two types of knowledge: on the one hand, knowledge making it possible to form rig-prospect combinations, sorted by order of preference[13] and, on the other hand, expertise making it possible to combine these different preferences in a programme that constituted the best compromise between all the preferences.

The first problem, the creation of rig-prospect combinations, did not present any major difficulties. It naturally required several successive conceptualizations, but the experts concerned were able to provide a large number of rules fairly easily. It is noteworthy that from this early stage they did not limit themselves to drilling rules, and that economic rules and preferences concerning contractors or the origin of rigs had to be introduced. Although expressed in the form of expertise, this set of rules was nevertheless characterized by considerable heterogeneity.

For the second problem, the preparation of a programme, the situation was totally different. *It was necessary to compensate for the almost total absence of expertise that could be formulated by experts.*

---

13 This order was calculated by means of a standard multi-attribute function; cf. Roy B. (1985).

4.1.3 Naval's planning skills: a prosthesis for human expertise

Invited to describe their planning practices, the experts could formulate nothing but broad principles. They were well aware of the main issues they usually defended during negotiations with the company's subsidiaries, yet they felt incapable of describing in the form of systematic know-how any particular method for putting together programmes. Moreover, even if they had been able to do so, their know-how would have been unusable since the practice of sorting rig-prospect combinations into an order of preference was totally new to them. They had only discovered and accepted the principle during the first meetings on the project. Planning expertise was, in other words, unidentifiable for them. Although they agreed that they were the best people in the company for the job, they left it up to the AI specialists to find a procedure for solving the problem – even if their experience was to be used afterwards in the application of the expert system.

The method selected was based on an iterative calculation of compromise by the propagation of constraints[14]. The expert system explored the different possible programmes, trying to reject the smallest number of recommendations based either on preferences for rig-prospect combinations, or on normal planning constraints relating to sequence, time or budget. Equipped with this method, the Naval system was unquestionably structured to fit somewhat uncomfortably into the main hypotheses of expert systems. Could it be considered as an expert system at all, in the sense of knowledge having been "captured" by the cognitician? The conceptualization of a compromise function and the search for the most efficient programme in terms of this function, were contributions of the AI team working on the project. They were not part of the knowledge of those persons designated as experts. It could not be said that this method "imitated" or "simulated" the reasoning held during negotiations, since the very concept of a qualified indicator of a quantity of compromise was totally foreign to the experts. *The programme produced by Naval therefore constituted a pure innovation in the context of expertise within the firm.*

The experts did, nevertheless, provide rules which were used by the system's designers. But these rules were used as elements in a method of reasoning which was completely different to existing practices, and it seemed that each new compromise was the result of an ad hoc line of reasoning. The most striking characteristic of the Naval system was therefore the obvious incorporation of a method unknown to the experts, in contrast with the situation in the TOTEM and Cornélius projects. In those cases, the experts' knowledge was transformed, but the system's reasoning remained simple logical deduction. What is it then that distinguishes

---

14 Descottes Y., Lacombe J.C. (1985). This method was first developed for the preparation of routings for machining operations, something which may seem surprising in view of the fact that Totem followed a completely different approach. It can, however, be explained by the absence of available know-know in the relevant context.

TOTEM's reasoning from the method of compromise used by Naval? One could, as is often the case, qualify the latter approach as "algorithmic" because it uses a construction principle that is mathematically "formulatable" (like the search for the maximum achieved by a function), whereas TOTEM relies on universal logic. This distinction is not, however, very fruitful and one could argue that the Naval method is also logic applied to a question of optimization, that is, a guided search for the best compromise. *It is thus less in the nature of the method than in its spirit that a significant change in the nature of the expertise can be found.* In TOTEM, the search is for the first routing that can be found by means of certain rules. This attitude, that we called "exploratory neutrality" is completely absent in Naval, since its approach is mainly one of combination and selection. The "strategist's" expertise appears here as being related essentially to combining and arranging, by means of continual connection between the formulation of planning objectives, on the one hand, and the selective formulation of the plan, on the other hand. The method used by Naval is therefore a mere prosthesis, a substitute for the absence of expertise, but it highlights the role of the continuous search that characterizes certain planning situations.

Is Naval still an expert system? The question warrants being asked, since it allows one to emphasize certain hypotheses by contrast. Naval presented the paradox of conforming to all, *but one*, of the typical characteristics of expert systems. It used a standard AI language (Lisp), it accepted the evolution of many of its rules, and it could explain the steps in its line of reasoning. The only point on which there was an unquestionable doubt as to it being an expert system has been discussed above – Naval diverged considerably from the reasoning of an identified human expert. But what is the significance of this criterion? What is its relevance in an industrial context?

We shall see that in the Naval case the question was particularly significant. In an organization expertise has to be acceptable and accepted, and without expertise that can be expressed formally, an expert may be more legitimate than a method.

4.1.4 Naval: a system for experts

The development of Naval posed significant problems in the identification of available knowledge. If there was clearly no expert capable of providing Naval with planning expertise, it was also because no actor had hitherto assumed this responsibility. An oilrig programme was obtained by the juxtaposition of all the subsidiaries' programmes. This juxtaposition was then examined by several committees that tried to reconcile the most obvious contradictions by way of negotiations. Committee meetings were generally prepared by specialists who tried to maintain the validity of the information discussed, or to examine the different subsidiaries' arguments put forward in favour of specific solutions to local problems.

It was mostly these specialists who acted as planning experts, but their main asset was the legitimacy that they had acquired with time and during certain crises, e.g. the power to intervene, to contest an economic or geological option, or to propose combinations and compromises. They therefore expected Naval to provide them with support and additional legitimacy in their task of coordinating the subsidiaries' exploration activities.

The experts were naturally the first users of Naval. They wanted the negotiators to accept it as a peacemaking tool since it assisted in the search for a compromise that took diverse constraints into account. From this point of view, Naval was undoubtedly a "system for experts". Whereas the routings produced by TOTEM were in many cases acceptable without the production planner's intervention, and whereas Cornélius' diagnoses were aimed at helping actors other than experts on the facility, the type of expertise used in Naval could not be separated from the experts; it remained totally dependent on their credibility and on the conviction with which they defended a plan proposed by the ES.

It is possible to recognize the artisan from the quality of his product, and the repairer or the physician from the restored order after his intervention, but how can the strategist be recognized? What can make one believe that his success today can be repeated tomorrow? Most of the unobtrusive actors in industrial life are, to varying degrees, strategists. Simply because of the complexity of their missions and the diverse negotiations that these imply, their expertise oscillates between several types of know-how. With the GESPI project[15], the fourth and last project examined here, we shall see how "combining know-how" can justifiably rely on "doing know-how".

4.2 The GESPI project: untangling the web in a large station

The station in which the GESPI project took place is certainly one of the most complex in the French railway network, and caters for over a thousand trains per day. Each train has to be assigned one of the 640 possible routes for entering the station, as well as one of the 30 platform tracks. Furthermore, the trains leaving the station have to use the same infrastructures as those entering it, and trains have to cross one another without causing incidents. This problem would be less serious if the only complication was that of a new "service" being introduced every semester, and if this same schedule applied every day during that period, without any modifications. But the quantitative complexity described above is compounded by an untold number of incidents which make such repetition impossible. Decisions concerning schedule changes can be made for diverse reasons, notably

---

15 See in Part Two the chapter "Gespi – Discovery of station traffic planners' expertise".

commercial, a few weeks or even a few days before their implementation, and tracks can be temporarily unavailable due to accidents or maintenance work. *It is therefore necessary to start every day with a new traffic schedule that takes into account the actual state of movements and infrastructures. This plan, or diagram of the assignment of trains to platform tracks, is prepared by a station planning department in which there are several employees.* It is then transmitted to the dispatchers who direct the traffic in real time, adhering to the plan as far as possible.

What is a valid plan? It is one which takes into account first and foremost a detailed description of the available facilities and the trains using them. It is also one which respects diverse constraints relative to either the routes or the platform tracks, e.g.

- *constraints concerning routes*: these consist, for example, of avoiding conflicts between trains; certain trains have to wait until their tracks are cleared before they can pull into the station;
- *constraints concerning platform tracks*: their sources vary. They may be imperative, if they are due to technical impossibilities (e.g. a platform that is not long enough for the train), but they may also be due to more functional needs (e.g. a commuter train must arrive in a zone equipped with date-stamp machines, and should always arrive on the same platform).

4.2.1 Station planners' expertise: between "combining know-how" and "doing know-how"

Out of all the projects that we analysed, GESPI is probably the one that changed the most during its development, precisely because the planners' expertise was perceived differently during the course of the project. The first approaches adopted by the project leaders were similar to those that we encountered in Naval, where the specialists tried to find a method that combined different constraints and resulted in a compromise programme. The planners' expertise was used to draw up a description of facilities in the station and constraints to be taken into account, and to index these constraints. However, the number of situations to be examined and the constraints described were not adequate selection criteria, and this approach was soon doomed to sink under the immensity of the number of possible combinations.

The system designers therefore progressively chose to structure the resolution of the problem by trying to break it down into more or less independent sub-problems. *They were thereby able to deploy the station planners' knowledge differently*, since the problem was broken down by means of rules which allowed the planners to structure their intervention. A first type of rule enabled them to detect potential traffic conflicts between trains and thereby to define groups of trains where a risk of conflict existed. A second significant type of rule consisted of distinguishing two classes of trains: on the one hand those which already had a platform track because they were a regular part of the service and, on the other hand, those which did not have one because they had been introduced afterwards. These two strategic rules made it possible to work out a programme in stages. For trains in the first category, it was necessary to check whether there was an available route on the relevant day, providing access to the assigned platform track without causing any traffic conflict. If no satisfactory route could be found, the initial assignment of a platform track had to be challenged and the train put into the second category. The system would then try to assign a platform track to trains in the second category by ensuring that there really was a conflict-free access route. Finally, if conflict seemed unavoidable, the system listed the trains concerned and these were then dealt with directly by the planners themselves.

This almost linear approach to the problem made extensive use of the planners' experience and know-how. It was their experience with the problem which enabled them to consider certain configurations as potentially more dangerous than others. For example, traffic conflicts are usually events which occur over a period of three to five minutes, their prediction is therefore limited by the inevitable uncertainty of train times, and some conflicts identified in advance can even disappear by themselves. The planners' experience similarly led them to consider that when trains already had platform tracks, it meant that they had already been screened once and that certain constraints had been disregarded purposely, whereas new trains without platform tracks could create severe difficulties.

Thus, the programme provided by GESPI partly resembled the production routing produced by TOTEM, in that it combined the main elements of "doing know-how" with the combinatorial skills required in the search for a compromise; in fact it had to examine systematically all the trains selected. Like TOTEM, it is not possible to say whether GESPI's solution was the best possible compromise; it was simply acceptable or not with respect to set constraints. GESPI used the experience and rules provided by the station planners, but contrary to all appearances, it did not imitate their reasoning.

4.2.2 GESPI: a different working procedure

On closer examination, it becomes clear that GESPI did not handle the same problem as the planners, even though they both produced the same plan. Of its initial combinatorial logic, the expert system retained a centralizing logic. It accumulated all the information characterizing a day's activity, before processing this simultaneously on the eve of that day. This method, intended to favour overall coherence, was not the one used by the planners who were used to dealing with modifications on an ongoing basis, as they received them. *Thus, a day for which there were many changes was worked on several times, and each new event concerning it led to a new plan. The planners were therefore thoroughly familiar with the details of these plans.* This difference in the working procedure created an unexpected paradox in the project's life: in spite of its rapidity, GESPI was less flexible than a station planner when it came to dealing with sudden last-minute modifications. In order to accommodate changes, the system had to reformulate all the programmes for the day, whereas the planner could usually make hasty alterations to his plan just before sending it out. The most recent changes to GESPI were aimed at loosening up data processing so as to maintain the greatest possible flexibility, and at giving the system a greater capacity to explain its reasoning. In the first experiments undertaken, the planners had difficulty in understanding why a train which they considered to be problem-free, was rejected outright by GESPI. At times it was perhaps done for a good reason.

Naval and GESPI provide examples of two different ways of modelling "combining know-how", but this difference is due as much to the nature of the problem as to its history in the firm. Trusting an actor with the task of drawing up an action plan that has to be a compromise between multiple objectives is tricky. If the actor becomes part of corporate life, he or she will be recognized as having certain "expertise", even if this is difficult to define. In the Naval case, such an actor had difficulty emerging, whereas with GESPI he was already there and had accumulated expertise that could be reused. From the moment this became clear, the aim of the projects was no longer the same, and the type of modelling adopted only emphasized this initial difference.

Our comparison of GESPI and Naval calls for another remark. In Naval it was an abstract model for the calculation of compromise that served as "combining know-how", whereas GESPI relied more on the experts' practices. The Naval approach provided the greatest capacity for evolution, since with GESPI a change in the evaluation of traffic constraints or of the assignment of platform tracks would necessitate considerable revision of the system. This difference was another result of the interconnection between knowledge and reasoning. *Naval was in a sense a quasi-universal tool for compromise, whilst GESPI adhered to the planner's specific situation. It is never easy to be highly relevant and extremely versatile*!

With "combining know-how" we have reached the end of our examination of the structure of the three main types of expertise. Our study was based on projects which all had fairly similar conceptions of the nature of expertise in a firm and ways of capturing it; it was the comparison between this conception and reality that made each project's story. Their differences and the obstacles encountered varied from one case to another. We were thus able to define the expertise involved in each of the projects, not by a study which grasped the nature of the expertise by means of observation or discussions with the relevant actors, but by revealing it through the vicissitudes of the modelling process, *a little like a pianist's talent might be discovered by contrast with a piano-playing machine.* Our analysis is not exhaustive, since it is limited to whatever can be brought to light in this type of comparison, but we shall complete it in subsequent chapters by looking at the history of the projects and their actors.

Before drawing our first conclusions, it should be noted that the analysis of expertise is also a fertile way of analysing the concept of technology.

4.3 Technology and expertise

Any activity includes a certain degree of "doing know-how", "understanding know-how", and "combining know-how", and it is important to understand the extent of each of these different components in order to gain adequate insight into the activity. One of the easiest mistakes to make is that of neglecting the role of "understanding know-how" and "combining know-how" in an activity, which is often the case in analyses of repetitive jobs. Such jobs are frequently seen as requiring only know-how relative to the tasks to be performed; yet this view ignores the fact that, in order for the execution of operations to proceed as planned, the environment must also conform to a plan. Countless minor changes could, however, occur in this environment (e.g. undelivered materials, machine failure, unexpected results), and a different approach would then be required in order to understand the problem and find a way of solving it or limiting its effects. Those tasks which appear to be the simplest can sometimes require great acumen and a sense of organization as soon as they have to be performed in fluctuating conditions. In contrast, a highly complex procedure or one considered as such because it requires the monitoring of hundreds of parameters, but whose laws and operating conditions are perfectly known and controlled, might require no more than "doing know-how".

Henceforth, when we talk of "technology", what implicit assumption are we making on the structure of the corresponding expertise? Can we talk of a repairer's technology as we would talk of the technology of tiling, for example? It is clear that in this respect the vocabulary is hopelessly inadequate. Several researchers have tried to characterize different types of technologies, and it may be useful to look at these classifications to see whether they correspond to different forms of expertise. Such an analysis is beyond the scope of this book, but our remark does suggest that the problems encountered by expert systems faced with the variety of expertise, highlight the vagueness of our most common notions.

But we must, at this point, conclude the results of our first journey into the heart of these projects.

5 A paradigm for multiple expertise?

Our trip through the world of expertise in practice has allowed us to draw several conclusions about expert systems and to define the overall characteristics of this type of project.

The first of these conclusions concerns the basic hypotheses of expert systems, which have to be transgressed. The necessity of doing so is due to two main facts which we shall now evoke again.

1 – The hypothesis of separation between knowledge representation and reasoning representation is relevant only to certain types of expertise.

This hypothesis is open to criticism from a philosophical point of view, for it decrees an arbitrary separation by denying certain reasoning the quality of being knowledge, even though this is the case for most mathematical models. Even if we exclude such a general debate, it is enough to note that the knowledge/reasoning separation is only operational under certain conditions, and notably when expertise can mainly be expressed as "doing know-how"; in other words, when it only mobilizes deductions and inductions from elementary knowledge without any particular articulation. Some authors[16] have pointed out the excessively restrictive nature of such a separation, yet they have not always seen that agreeing to forego such a construction axiom strips expert systems of one of the elements most commonly put forward to denote the specificity of the discipline.

One of the advantages of the separation between knowledge and reasoning has been the possibility of updating the system easily, at the same pace as the knowledge itself, without changing the reasoning. If, however, this separation is not possible – and we have seen that it is a borderline case – new knowledge may well give rise to the need to reorganize the system. The possibilities for updating an expert system cannot be determined in advance, they are evaluated differently for each project.

Finally, the idea of a knowledge/reasoning separation can lead to practices which hinder the project. The argument is simple but logically sound: if knowledge and reasoning are perfectly separable, it is possible to capture an expert's knowledge without him knowing anything of the corresponding reasoning. What is then the point of teaching the expert about the principles of an inference engine, since he need only reveal his knowledge? The inherent danger is clear: the "expert" is also, in most cases, the user of the system, and his own knowledge would then be available to him in an opaque and unmanageable form. The fact that the system "explains" its decisions in certain cases, can reduce the phenomenon without making it disappear. The explanation must also be explicable!

---

16   Bonnet A. et al. (1986).

5 A paradigm for multiple expertise?   55

2 – Expert systems neither imitate reasoning nor capture knowledge; they can only function if they transform it.

We have already mentioned, in the preceding chapter, the terms denoting the way in which expert systems are built using experts' know-how; "collect", "imitate", "capture", "reproduce" are the most commonly used verbs. These words are vague and allusive, but they all evoke the same type of process, that of expertise flowing naturally from the expert's brain into the computer's memory. The image is not only totally inadequate, it is also unfair towards the system designers, for the modelling performed by them is far from being a mere exercise in reproduction. On the contrary, it is unquestionably one of "transformation", comprising at least three aspects of relative and variable significance, depending on the situation: the selection of existing knowledge, the restructuring of this knowledge, and the incorporation of new knowledge.

– *Selection*: everything that the expert knows is perhaps not relevant to the project or may not be suitable for putting into a computerized form (e.g. the description of a station chosen in GESPI was simpler than one given spontaneously by a station planner, while TOTEM was only possible on certain manufacturing activities).
– *Structuring/restructuring*: The formulation of expertise is not an operation of pure translation. It consists of finding new concepts which can be manipulated more easily or which are more relevant to the problem that has to be solved. Thus, in Naval, a first version of the project only dealt with oilrigs and wells to be drilled. A few months later it was clear that this elementary structure was inadequate and that it was necessary to create the object "contract" to express the commercial relationship linking an oilrig to a company during a particular period. In retrospect, this modification can easily be explained. Without the formalization of the object "contract", it was always necessary to talk of the compatibility between a rig and planned exploration. But for a given period the rig may or may not have been available. Since this variable influenced the desired compatibility, it was necessary to distinguish each of the periods in which the commercial state of the rig changed. The complexity of the representation then became inextricable, unless the contracts were considered as objects in their own right and a definition made, not of the compatibility between a rig and a prospect, but between a contract and a prospect. A similar conceptualization is also found in expertise that is easier to model, even if its modalities are simple, such as the hierarchical organization of stages or the separation of distinct "areas of know-how" within the same expertise.
– *Incorporation of new knowledge*: This last aspect is the one that is the most clearly opposed to the idea of "imitation" or "collection" of experts' know-how. We have encountered several examples of it in the projects analysed; for example, the compromise model in Naval, the calculation of dimensions of the basic metal for routing in TOTEM, or the model of the descriptions of interac-

tion between components of the same equipment in Cornélius. In all cases, the designer contributed not a language, but new expertise which was useful in the resolution of the problem and contributed considerably to endowing the final system with some of its most salient features.

These conclusions originate in one basic fact: the concept of "knowledge" is far too vast for the vocabulary of expert systems, or that of knowledge-based systems, to refer to a single methodology. By taking for their field of application expertise as diverse as that of the artisan, the repairer or the strategist, the expert system leaders had to choose between two strategies: that of maintaining dogmas in inappropriate contexts, or that of renouncing these dogmas in practice, in favour of a more flexible approach. The second option seems to have prevailed in the field, even if it was not enough to guarantee the relevance of the project. It is noteworthy, however, that at least one hypothesis was left intact: that which leaves it up to the expert to validate the results of the system which has been developed. We shall see that it plays an important role in the dynamics of all the projects.

If expert systems cannot find in a set of specific hypotheses the substance of their approach, what exactly are they? We are of the opinion that in losing an identity which is too limited, expert systems, paradoxically, become more meaningful and find their roots. Through them, artificial intelligence has invented a new figure of rationalization; it extends the project of automating intellectual tasks by examining the structure of know-how and in particular of that which can be computerized.

Despite its banality, the word automation expresses the reality and the variety of the expert systems that we encountered far more accurately than do the basic hypotheses of the discipline. At least it refers to a historical experience which showed us that the process can never be reduced to mere imitation, that it demands the production of new knowledge, that it is not always synonymous with efficiency, and finally that it results in significant transformations in the collective conditions of the action. Automating expertise is first and foremost trying to identify it, it is also changing it, and it is, in particular, understanding how these two operations are likely to be accepted by the actors that it involves in an uncertain metamorphosis. That is what we shall try to do by looking once more at the course and the actors of the projects studied.

Chapter 3
Life of Expertise and Metamorphosis of actors
Birth, crises and development of expert systems

If all knowledge could be drawn from experts like water from a spring, an expert system project would be a mere episode in corporate life, just another operation in the production process. Yet nothing in these projects is that simple. While developing expert systems implies a process of exploration and discovery of certain actors' expertise, their impact depends mainly on the reformulation of this expertise and on the new relationships it spawns. *The fact that the effective survival of an expert system is only one of the possible results of this reformulation is all too often forgotten.*

It is nevertheless easier to characterize expertise modelled in each project, than to analyse the complex paths that lead to its reformulation. Each case history has its singularities and is situated in a particular context. This difficulty is sometimes overcome by comparing the actual development of a project with a specific model or methodology. One can, for example, define certain essential steps to be taken or state the importance of detailing specifications before implementing a project. One can similarly emphasize the importance of maintaining a regular pace of development[1] or insist on the necessity to efficiently manage the key moments when the project has to be redefined in the light of new insight into the issues involved. In the cases examined here, the role of these notions differs from one project to another so that discussions on methodology are often too abstract to be relevant. In fact it is difficult to understand the course of a project without relating the content of the expertise deployed to the history of the actors involved in it[2].

General descriptions of the actors in an expert system project usually evoke four main characters: *the experts, the specialists in the discipline, the users, and the project leaders*. This categorization is, however, somewhat theoretical and is seldom observed in real situations. Some actors have several roles simultaneously – as was the case in TOTEM, where the project leader also acted as a specialist – and it may be difficult to identify them accurately. Throughout the Cornélius and Naval projects the definition of users was more a subject of permanent debate than

---

1 Grundstein M., de Bonnières P. (1987).
2 O'Leary D.E., Turban E. (1987); Stamper R. (1988).

an initial option. As for experts, they are not just walking encyclopaedias waiting for their knowledge to be captured. It quickly becomes clear that their expertise is characterized by their mission, and is closely related to their relations with other actors.

One therefore has to look behind this collection of individuals, at the industrial situations in which the projects were born, and at the specific organizational contexts and stakes which gave meaning to each player's action. We shall first analyse the conditions in which the expert system projects were initiated, and discover that they were characterized by certain features typical of all innovation and by other aspects peculiar to this kind of project. We shall then look at the types of actor that served as experts, which will enable us to situate their traditional position in the industrial world. Finally, we shall attempt to understand the changes which took place during the course of the projects, by highlighting *the metamorphosis of actors that correlates with the recomposition of expertise throughout the project.* This process of metamorphosis is always present although to a greater or lesser degree, depending on the case. For a better understanding of the phenomenon we shall outline the theory underlying it, and so identify how it differs from more conventional models of organizational change.

1 Birth of the project: myths and innovators

In spite of the retrospective view that can be taken of expert systems, and the considerable number of publications on ES "applications", the context in which these projects were born is usually either ignored or simply taken for granted. *The remedy selected is therefore often considered to be a symptom indicating the cause of the problem*; for example, a diagnostic support system is chosen because of problems in diagnosing failure. But to understand the birth of these projects far more information is required: What was the origin of the problems? What were the alternatives? Who investigated the problem and opted for the solution? Answers to these questions are not merely anecdotal; it is in the manner in which expert system projects are born and organized that a number of realities are revealed, including the image of expert systems as conveyed by the different actors, the dynamics of expertise, and the state of each actor's knowledge.

The first remark that can be made in this respect concerns the exceptional nature of the expert system adventure. Although the cases we studied dated from 1985 onwards, when expert systems were not yet commonplace in industry, the media coverage they enjoyed was far greater than that devoted to any other type of computerization. This was undoubtedly because expert systems had (and still have) a high-tech, somewhat futuristic, image, with the reputation of being highly spe-

1 Birth of the project: myths and innovators

cialized. As a result two main features emerged which characterized the birth of the projects: *the mythologization of an industrial problem, and the role of actors outside the usual structure of the firm.*

1.1 Mythologization of an industrial problem

All the projects we studied were based on perfectly real industrial problems, yet the actors that had to promote the expert system solution were not familiar enough with them to be able to articulate sound arguments in support of their choice. At the outset, the problem was presented in more or less abstract terms which *a priori* corresponded to the basic properties typically associated with expert systems. It is this process of abstraction and interpretation that we call *mythologization*, precisely because it deviates from reality to promote a particular course of action. We shall see that its presence varied from one project to another.

1.1.1 Discovery and reinvention of production routing

We shall focus on TOTEM first, because in it the mythologization of the problem was reduced, without being absent altogether. The project leader already had experience with a similar project, although one that had taken place in a different industrial context, that of a factory producing aluminium sheets. Even though he had worked with expert systems, he was unfamiliar with an industrial tradition that proved to be far less suited to the elaboration of systematic work procedures than the aluminium industry.

The initial objective of the TOTEM project was simple. *The variety of alloys and of products to manufacture had increased so fast that it threatened production planners with an excessive workload and resulting errors.* The automation of production routings seemed therefore to be a possible solution. It nevertheless necessitated an understanding not only of how routings were prepared, but also of the developments that had led to their present form. Hence, TOTEM was impossible without careful thought on the initial contents of routings and on their role in the production process. It was also necessary to decide exactly when routings could be handled by the new system, for automation can lead to overkill. Was it necessary, for example, to describe in more detail the list of operations to be carried out in a given production process? Was it necessary to involve – as different sources of expertise – not only the production planners but also the shop supervisors and operators? Was it preferable to limit the scope of the project by only producing routings similar to existing ones? These questions were not easy to formulate at the start of the project and it was during the effort at revealing and reformulating expertise that they emerged. But greater clarity also highlighted the inconsistencies or ambiguities of certain routings, that workshop personnel had become used to correcting themselves, and the fact that some procedures had become outdated without this being noticed. As it developed, automation paved the way for several unexpected operations, such as the updating or improvement of the coherence of basic axioms in routings, and the enhancement of these by the inclusion of new types of information. *Thus, between the routing that was imagined in abstract terms at the start of the project, the one that was effectively produced manually, and the one that was finally selected in TOTEM, there was a continuous process of invention fed by the modelling process.*

1.1.2 An inappropriate perception of the job

The creative process encountered more difficulties in the Cornélius and GESPI cases than in TOTEM, and in GESPI several reconceptions of the project were necessary before a final form was reached. The first version of the project was aimed at automating the dispatchers' work in real time. The choice of a route or a platform track was to have been proposed by the system, just before the dispatchers took their decision. This idea, deemed unrealistic in terms of the constraints weighing on a decision in real time (necessity for a rapid response, perfect reliability and continuous updating of the system's information), was rapidly abandoned. But perhaps it was also the direct link with the dispatchers' actual tasks that modified the original idea to deal with this activity first. By rather tackling the problem of the plan assigning platform tracks, the project leaders were dealing with an activity which did not belong to the dispatchers, but which did nevertheless have the task of guiding them by means of a preliminary plan.

The final version of GESPI was only developed several steps later. Under the guidance of the specialists leading the project, the planning problem was first posed as a combinatorial one, before this conception was abandoned. Of interest is the fact that this early approach was judged acceptable by certain former planners and by the executive team managing the project, and was only challenged when difficulties were encountered and when the planning team manager intervened. This shows how, in retrospect, or with a perception that is too far removed from the real mechanisms of action, there is a tendency to adopt attractive and slightly mythical representations of the problems involved.

After several months of work initial conceptions faded into the background as attention was focussed on more limited problems and notably on the groups of trains subject to traffic conflict. A similar evolution also took place in the representation of the traffic planners' work. The initial idea that the planners drew up an original plan every day, was abandoned when it became clear that they tried rather to preserve a standard plan for as long as unacceptable conflicts did not appear. Significantly, the team working on GESPI dealt with these conceptual transformations in a skillful manner throughout the project and willingly reviewed the

1 Birth of the project: myths and innovators   61

initial approach. *This type of reconceptualization is neither a matter of delving more deeply into a particular problem, nor an incremental learning process, but a radical change of perception that springs from the discovery of the mythical nature of the initial vision of the project.*

In the development of Cornélius there was no such distinct reconception of the project. Nevertheless, an apparently reasonable notion such as supporting failure diagnosis followed a sometimes chaotic path before any clear definition was made of exactly which diagnosis of which failures could be supported. It was in fact the meaning of the concept of support which fluctuated the most throughout the project. This was due to the fact that, in general, *the actor who needs assistance is in the best position to identify the assistance that will be most useful to him or her*. Support must be intelligible, both in its contents and in the way in which it intervenes. We shall see later that the stakes involved in the Cornélius project were not important enough for it to be developed to a successful conclusion.

1.1.3 Computerizing the ideal compromise

It was in Naval that the most significant mythologization of the initial question was to be found. The concept chosen at the start of the project was based almost exclusively on a mental figure: *the interpretation of negotiation between several partners, as a calculation of the best compromise*. The distance between this figure and reality was such that the discourse initially held by the project leaders had difficulty in giving even metaphorical substance to the project. The firm's AI specialists wanted to develop their skills in "symbolic calculation, the only way of reaching a compromise"; some practitioners saw in the project the hope of taking into account the subtleties and unknown factors in tricky negotiations between head office and the subsidiaries; others were more concerned with rationalizing what appeared to them as a set of decisions taken on an ad hoc basis, without any overall coherence. Clearly, only the futuristic aura surrounding artificial intelligence could give such a project any credibility. Without the impression that this was an exceptional type of computerization, capable of a new type of calculation, the project would never have been supported by management, in spite of the undeniable problems created by the scheduling of rigs. *This gap with reality, despite the intense efforts of the two experts concerned, was never totally closed.* As the tool took shape the aura disappeared, and the context in which the experts worked appeared more clearly during the modelling process. Thus, the image of Naval changed progressively from that of a powerful tool for reconciling contradictory interests, to the more modest and realistic one of a safety-barrier to avoid certain practices during complex negotiations; for example, the use of inadequate arguments, the hasty adoption of the first solution on which there was consensus, the unbalanced attention given to conflicts (the most complex ones did not always receive the most attention). But even if this analysis made the tool less idealistic in theory, it still had to be acceptable in practice.

1.2 Innovators: specialists or interveners?

The mythologization mechanism described above indicates the reality of the uneven distribution of expertise, at the start of projects, between the promoters of rationalization who were barely familiar with the problems, and the experts initially designated as such. It is also obvious that the projects were not conceived by the actors most directly concerned by the problem. The reason was no doubt the natural passivity of those confronted on a daily basis by a reality to which they adapt as best they can, or the fact that projects such as these can only be launched by staff on a certain hierarchical level. But perhaps we should also add a third hypothesis relating to the problem of certain actors' expertise. Those who know an industrial problem well, because they have experience with it, also know that there are several ways to solve or transform it; they know only too well that today's formulation of the problem is perhaps not tomorrow's, or that the issues can shift. The results that can be expected from an expert system, for example, are often relativized and compared with all the other possible sources of progress. *The production planners in TOTEM, the station planners in GESPI, and the operators in Cornélius, would undoubtedly have given as much support to other projects aimed at helping them in different ways.*

Who then are the promoters of these systems? We encountered two main types: the AI specialist and the "intervener" responsible for an innovation process. They were, moreover, sometimes present together on the same project.

The AI specialist can act as a "technical" manager or even as a project leader, as in the GESPI case. His role was even more important in Naval, where he appeared as the project initiator, the one who had managed to gain acceptance for it and who therefore had a right to intervene. In the two other cases studied, the specialist was either absent – when the acquisition of a software program was considered as an adequate investment – or acted as a technical advisor who was called upon when needed, but had no responsibility for the project. The AI specialist's role is never insignificant in the life of an ES project since these projects tend to be considered as experiments for learning more about artificial intelligence. They are the scene of difficult choices and are generators of potential conflict, for discovering a technology does not necessarily mean its effective implementation, notably if progress is achieved by other means. The AI specialists may well fear a hasty and unfavourable judgement of the technology, when the problem to which it is applied has been formulated incorrectly. The ambiguity that has been maintained between a general educative goal, on the one hand, and the concern for immediate efficiency vis-à-vis a specific problem, on the other, is probably the main threat to this type of project and one which places AI specialists in an uncomfortable position. Such ambiguity is probably inevitable in innovative projects, but it is up to other actors to deal with it at key moments during the course of the project.

We shall use the word "*intervener*" to denote those actors who, although removed from particular problems, become directly involved in analysing and dealing with them during the course of the project. They may be the project leader him- or herself, or experts who are there to support the innovation. It was in TOTEM that this role was the most obvious. From the outset the project was handed to an engineer recruited especially for this purpose and who, besides being an AI specialist, was thoroughly familiar with the operation of a metallurgical factory. He therefore had a very specific perception of the issues from the first stages of the project. On the one hand, the automation of production routing was not perceived in isolation, or for its own sake; the TOTEM project was situated within the overall framework of problems encountered in production management and considered as a means towards more general computerization in the firm. On the other hand, the conception of the project's organizational and social impact was constantly reviewed, which in turn influenced the procedure adopted for it. Thus, the success of the project was progressively linked to the effective evolution of the workshops' functioning.

A key player like the TOTEM project leader was not clearly apparent in Cornélius, although the project did benefit from the intervention of a support service for the introduction of new technologies. It was a member of this unit that led the work group responsible for developing the system and clarifying the multiple objectives which could be assigned to a diagnostic support software program. This type of actor is to be found in most innovation[3] but one feature of expert system projects intensifies the need for "interveners", i.e. *the complexity and the duration of relationships that have to be established with the experts*. The management of a project's basic strategy cannot be dissociated from a capacity to involve the experts, which in turn demands dual legitimacy that cannot always be found in the same person. All industrial contexts do not permit the intrusion of such actors in the same way and do not necessarily have the right people available.

If a project is conceived according to a mythical model, it rapidly has to face the test of reality. One, and not the least, aspect of this reality relates to those designated as experts. Who are they? And what guarantee is there that they really are experts? The mechanisms which make it possible to designate an expert as such are clearly complex and varied. A university degree is all that is needed to recognize a lawyer or a doctor, and in our daily life we often leave it up to institutional systems to decide who are experts and who are not. *But corporate functioning does not lend itself readily to such a limited definition of the word.*

---

3 Alter N. (1985); Sainsaulieu R. (1989).

2 Experts in organizations: nature of expertise and position of the actors

The word "expert" is not nearly as commonly used in corporate life as "specialist", and if we examine the prevailing theoretical syntheses[4] in which the main types of enterprise or organization are listed, we notice that this notion can denote very different kinds of actors. One could qualify as an expert a methods engineer in an industrial firm, a specialist such as a doctor or a geologist, or a consultant. Yet, the nature of this expertise and the way in which it is utilized and recognized, are very different in each case. A methods engineer in a factory is often unable to use references as specific as those of a cardiologist to define his or her field of expertise or mode of action. *There is, moreover, no reason why the simplest operator cannot be called an expert* (starting with Taylor, a long tradition of analysts has tried to define workers' expertise or that of operators in the field) but the concept of expertise would then be endowed with limitless ubiquity – either everybody is an expert or nobody is one. It is therefore necessary, in each case, to refer to the contents of the actors' expertise, to their history, to the value that is attributed to them, and to the way in which these mechanisms define perceptions, roles, or positions; in short, to the way in which they acquire meaning in a network of relationships.

After all, corporate history is also that of occupations which come into being and disappear, of forms of expertise which were unaccepted yesterday and are recognized today. It may seem that the function makes the system, and that the need makes the expert, but this need still has to make the existence of the expertise accepted. Who would refuse an expert for winning the state lottery! But something which is acceptable for horse racing is no longer so in the face of pure chance; the idea of a forecaster of lotteries is scarcely imaginable! The example is of course a caricature, but at the start of the century, who knew what a marketing expert was? The identification of an expert implies, above all, a definition of the relationship between the validity of expertise and its acceptability in collective action. The way in which this expertise can be shared, and therefore transmitted and diffused, must similarly be defined. Today there are fewer public writers than in the past, not because we no longer need to write but, on the contrary, because this expertise has been popularized.

To understand the developments implicit in the automation of certain types of expertise, we have to look at their history, at the origins their *validity*, their *acceptability* and their distribution. This is what we shall do now by sketching the profile of those players who acted as experts in the projects studied.

---

4   Mintzberg H. (1979).

2.1 Workshop planners: a trade born with Taylorism

We know that the idea of a "work scheduling" office was born with the experiments carried out by Taylor and the engineers who were part of the movement he started. The general idea was clear, but could include a large variety of roles and trades. The choice of tools, the definition of steps to be carried out, the setting of machine parameters, the specification of raw materials, the design of control procedures and the evaluation of production costs, were all missions that were to end up as the responsibility of these offices. They all evoke a particular, productive system, *that in which the daily functioning includes decisions taken to adapt the manufacturing process to suit the desired products*. Thus, the planner's office has no real use in a fully automated system; in this case a set-up man and one or two operators can generally supervise the facility, and the pace of production can be defined in its design. The planner's work makes sense when production is the result of the "appropriate" implementation of certain means in varying ways, and where this implementation also demands an overall view of available resources and processes. This mission implies the utilization of technical, economic and social data. The choice of a tool or a machine can be guided as much by the purely physical feasibility of the operation to be carried out, as by the "speed" with which it is performed, or the quality of the available operator. The production planner can then be described as an "orchestrator", dealing with technical parameters and notions of productivity or cost effectiveness alike, as well as with human skills.

Planners' expertise is therefore multiple and intermediary, and can become so specific downstream as to resemble that of a workshop supervisor or an operator, positions which planners have often occupied themselves. It can also become broad enough upstream to include essential data on the procedures at stake, much like commercial engineering. Within this continuum, the planner's job is reinvented specifically for each context; the nature of technical procedures, the economic constraints, and the historically constituted modes of interaction between other actors, will all contribute towards a definition of his or her mission at a given time. *There are, for example, hardly any workshop planners in the chemical industries* where procedures are implicitly defined in the physical process itself. Sometimes the figure of the planner can merge with other collective actors. At the end of the 1970s, in a firm specializing in the cold rolling of special steels, the variety of markets and products led to the inclusion of production planners in teams managing product lines[5], so as to maintain close communication, and even partial versatility, between production planners, strategic planners and commercial engineers.

Finally, in certain cases supervisors and operators have to complete or even modify the planners' directions in order to take into account the real state of the equipment, products or workers. The planners will then appear more like keepers of a thesaurus of general standards and procedures, who maintain and update

---

5 Hatchuel A., Molet H. (1983).

this heritage. For they also have to adapt to the flow of innovations introduced by the research or design departments, such as new technical or commercial requirements, or new materials that destabilize or complicate existing routings. The planners are then themselves dependent on other expertise which they must translate or interpret in the routings they prepare. If their expertise tends towards "doing know-how", as was shown in the preceding chapter, it is formed in the same way as transition know-how, somewhere between the abstract models of the research department and the contingencies of production action itself. *The planner thus embodies a specific mode of representation and operation of the productive system*; it was this specific mode that the designers of the expert system had to explore and that they also helped to change.

2.2 Station traffic planners: logic of the station

Station planners, designated as the experts in the GESPI project, are much like production planners in industry. They do not intervene directly in the dispatching of trains, but "prepare" the work of the dispatchers who control the tracks leading into the station. They are also dependent, upstream, on the timetables which commit them commercially vis-à-vis the passengers, and on the technicians who assign tracks to the trains or define the movement of equipment in various places in the station.

But, whereas there is a strong resemblance between the formal aspects of these two occupations, the productive systems in which they are involved differ considerably. The effect is immediately visible when one examines the nature of the expertise in each case. In the absence of incidents or disturbances, station planners can limit themselves to using the same plan every day, as long as the timetable is the same. *In this case their intervention does not create a standard for work, it merely controls its coherence*, and amends it if necessary or shows its impossibility. They choose neither trains nor arrival times nor some of the platform tracks, but have to draw up a viable plan of activity in the station as soon as one of these terms changes. Their technical knowledge is easily identifiable and is in this sense similar to that of production planners (e.g. detailed knowledge of the station's infrastructures, usable routes, trains, rules, etc.). But whilst the expertise peculiar to their specific job is situated well beyond this knowledge, it constitutes the "core" of production planners' expertise. Station planners have to do something that is constantly renewed, something that is difficult to describe and to evaluate. What is a "good" station planner? How can he or she be recognized? It seems that only peer evaluation is possible, since each situation is unique, where an adequate combination of platform tracks and routes has to be found. This different position of the "core" of knowledge and of its mode of validation in the two cases, i.e. that of production planners and that of station planners, is essential for an understanding of these two actors; in any case it shows very clearly that the nature of their occupations is not the same. When production planners solve a technical problem, they know that they can keep the solution and reuse it as it is, or, better still, try to reproduce the method exactly to get the same result again. In contrast, station planners can reason only by analogy and association; for example, a certain combination of disturbances resembles another similar situation, or a certain train must always be placed first to facilitate a rapid solution, and so forth.

Station planners explore the logic of the station and its limits daily, but it is a logic which they sense rather than control. An old observation in operational research is that when faced with combinatorics, one lives by one's wits. Station planners, through their difficulties, can reveal an accumulation of constraints which have become unbearable, or certain inconsistencies between schedules and facilities. But whereas production planners can show that a method does not work as planned, station planners themselves are the sole gauge in assessing the difficulty.

2.3 Maintenance specialists: organizing expertise hierarchically

Maintenance specialists can certainly be considered as experts. Their predecessors were set-up men and their expertise is related to identifiable devices whose secrets they know. In contrast with station planners and production planners, the nature of their knowledge seems less dependent on a set of organizational relationships. This autonomy is, however, relative. Maintenance fitters must either share a part of their knowledge or be condemned to intervening in the most trivial problems, at the slightest sign of failure. Pushed to its extreme, this logic would kill machining; it is never possible for very long. Thus, as soon as equipment shows a high degree of sophistication, a problem of grading expertise and modes of intervention is inevitably posed. Industry has for a long time been confronted with this question which led certain sectors (e.g. armaments, electronics, space) to try very strict coding of maintenance levels.

However, each time a new technical facility is acquired, this typology of expertise and actors has to be reconstructed ad hoc. Questions arise as to whether divisions should be made according to the technology concerned (mechanical, electrical, pneumatic, etc.), or to functional sub-systems, or whether graded general levels should be distinguished. The medical field, because it is in a sense a maintenance discipline, offers a wide range of situations where the attribution of expertise takes place through a network of actors responsible for more or less well defined zones of intervention. It is moreover possible to observe the problems inherent in this attribution, such as the compartmentalization of specialities, the position of general practitioners vis-à-vis specialists, and the complexity and ambiguity of hierarchical relationships between doctors and nurses owing to the increase in the latter's knowledge.

Maintenance fitters also have multiple identities. When they ensure reliability they may well wonder how anyone can manage without them; they are then ideas people who may see a breakdown as a source of expertise. When they act as repairers they have to put matters right as efficiently and as quickly as possible. A relationship develops between them and the users of a facility (operators or methods engineers), which may be antagonistic or co-operative, depending on each actor's appreciation of the other's expertise and method of intervention. It is thus always possible to read the tension inherent in any technical system, through the expertise and practices of the maintenance fitters. The complexity of highly automated facilities and the importance of economic stakes related to the continuous use of this equipment, make maintenance a crucial activity which tends to enter into the preoccupations of all the actors. The involvement of operators in preventive maintenance missions or in certain corrective operations is often sought. At the extreme, if automation is total, the frontiers become blurred and the daily running of facilities is no longer very different from a maintenance activity.

2.4 Scheduling oilrigs: experts or negotiators?

With the expert in the Naval project we are looking at a different type of individual to those shaped by industrial history, as were the production planner and maintenance specialist, while the context of the project was completely different to that of the planners in the GESPI project. We are entering into the sphere of senior corporate management logistics, where several logics are at play. Although the choice of rigs in the Naval case was dependent on technicians, it also involved the company's exploration policies and relations between the parent company and its subsidiaries. Consequently, the schedule to be drawn up could not merely be based on a decision taken by an expert or a manager without due regard to the subsidiaries' arguments; it necessarily implied complex negotiations. The experts in Naval did effectively play a significant role in the preparation of these negotiations, but this role was largely dependent on the relationship that they managed to build up with the different subsidiaries concerned, and their experience was manifested in the complex choices they made. They had sound experience concerning oilrigs, extensive knowledge in geology, and were used to commercial and financial negotiations, but their expertise was only acceptable in so far as they themselves were acceptable interveners.

They had gained recognition by head office and the subsidiaries during the serious crises on the oilrig market, when they had managed to ease certain tensions and organize consultation between the different parties. Expertise thus had an original form here, *it consisted of the assumption that know-how, which also drew its worth from the expert's personal history, existed*. This type of actor provides us with an interesting model, since it allows us to demonstrate how a stock of know-how and an individual heritage of relationships combine to form expertise. Both factors are obviously always present, but to varying degrees; what was striking in the Naval case was that they were equally important.

Thus, the four types of expert mobilized in the projects studied existed differently. In order to understand them we needed to characterize the way in which their expertise was constituted, as well as the networks of relationships in which it became active. It was the balance between these two factors that the projects were to shift.

3 Dynamics of the projects: multiple lines of transformation

Expert system projects aim mainly at the automation of certain experts' know-how. But the changes they provoke are not the result of this objective alone, they also originate in the actual process of the project underway, which comprises two active lines of transformation:

– The first is due to the involvement of the specialists and interveners in a long-term relationship with the experts. For the former the experience is an intense one of learning and discovering the world of the experts, their know-how, and their relationships. For the latter, it is one of learning about relationships of exchange, about explanation and discussion outside of the normal working context. This stage can therefore not be seen as a period of research, and *the course of the project is in itself a process of organizational transformation*; whatever happens and exists in that context is an innovation, whether the project is successfully concluded or not.
– The second line of transformation follows the life of the automaton (the software). It spawns challenges and problems, and becomes the pivot of a series of reactions, adjustments, or projects which will result either in its rejection, its transformation (the project is launched again) or its integration, i.e. the software is granted the right to be considered, for example, as a tool, method, technology, aid or means (it does not matter what qualifier is used). This is when the essential stakes in this integration take shape, i.e. those that finally prevail, that were not necessarily perceived at the outset, but that became manifest during the life of the project.

Naturally, these two lines of transformation cross all the time, even if the second one tends to rapidly dominate the first when the expert system is implemented. But they both constitute *a movement*, a series of changes whose effects will, depending on the case, take on a form identifiable by both an observer and the individuals concerned.

In considering these questions we cannot be content just to refer to conventional trends in the study of change in organizations; a complementary perspective has to be adopted in order to gain insight into the substance and dynamics of this type of transformation. Before outlining some theoretical elements of the perspective, essentially by way of a discussion on "strategic" analyses of change in organizations, we shall examine the main features of the movement which took place in the different projects.

3.1 TOTEM: the basis of a new engineering science

In 1990 TOTEM is over three years old and has been operational – in other words, preparing production routings -for about eighteen months. In that time its field of operation has expanded, but it is not yet the pivot of a broad system of production management that its project leader originally envisaged. Nevertheless, the exceptionally detailed work carried out on the rolling and wire drawing workshops' routings has had several consequences.

Determining the exact nature and role of "manual" routings meant re-examining the assignment of responsibilities and competence between planners and supervisors. In one case the idea of an expert system was abandoned entirely since, in that particular workshop, routing consisted of no more than a general outline of the work to be performed. Technical choices were made on an ad hoc basis in the workshop during the production process. The automation of such route sheets would have resulted either in a model of little use, or in the revision of barely stabilized techniques, presenting a number of situations that were too varied and unstable to be listed. In such a context, planning was limited to administrative support. By contrast, in most of the other workshops TOTEM was carried through to completion. But, once again, the expert system's routing is not identical to manual routing.

For the production planners, expressing their choices in the form of rules linked by a conceptual structure did not come naturally. As a result, the knowledge modelling phase took longer than expected despite the unquestionable experience of the project designer. In order to understand the reasons behind these difficulties, it is necessary to look once again at the differences between a manual routing and one generated by TOTEM, bearing in mind that each TOTEM routing is the system's solution for a specific commercial order.

3 Dynamics of the projects: multiple lines of transformation

Even if a human learning process can be represented as a series of successive steps, going from simple and repetitive routings towards more and more complex or specific ones, this is certainly not the case with TOTEM. In other words, in contrast to the genetic mode which, according to certain theoreticians like Piaget, is the mode of human learning, TOTEM uses a hierarchical structure of concepts and knowledge in which all routings, from the simplest to the most complex, are processed in the same way.

The project leaders were therefore forced to admit that, despite it being easy to operate and even though it had been built largely with their knowledge, TOTEM could not be maintained, and therefore made viable, by the production planners alone, at least not with their level of knowledge. *This is all the more interesting in view of the fact that TOTEM did not know as much as they did*! A fairly large proportion of the routings are still prepared manually today (10 to 20%, depending on the workshop) precisely because they are too singular or too unstable for the expert system.

TOTEM is therefore not only a routing machine, it is a bank of knowledge and concepts which acts as an engineering science for many production processes. The production planners' expertise has thus been reformulated into two areas:

– complex and barely formalized know-how, related to innovations underway, to orders that are too singular, or to certain peculiarities of the operators and the machines; this expertise remains the production planners' prerogative;
– expertise produced by the production planners' experience but reformalized (or axiomized) and completed for TOTEM; it relates more to an abstract vision of metallurgial transformations.

The reformulation of expertise had two main unexpected consequences: the appearance of a methods engineer and a new way of planning work.

Industrial life in the workshops in which TOTEM was implemented was of a somewhat traditional nature. The engineers were mainly to be found in the research department or in senior management, and production planning was seen less as a locus for the creation of production methods than as one for their efficient application. During preceding years the increasing complexity of these activities had forced the firm to pay closer attention to its production structures and essentially to problems relating to stock and quality control. A clear definition of working procedures or protocol was no longer sufficient, and a better way of managing the real production processes had to be found. TOTEM was born of this evolution, and it strengthened the tendency considerably. The expert system demonstrated that new manufacturing expertise existed, and thereby offered appropriate mediation for the introduction of a new actor, a methods engineer. It seemed natural that the maintenance of the system and responsibility for production planning be given to him straight away.

Besides this immediate development, TOTEM had other more long-term consequences. The effective and lasting survival of the expert system necessarily implied a permanent transfer of new knowledge acquired by the production planners, as soon as it was able to be formalized. This meant that their mode of manually preparing routings would be modified. But, although the methods engineer was to be the only keeper of TOTEM and the only person authorized to update it in any significant way, it was not possible to do without the active participation of the planners. Excluding them would lead to considerable discrepancy between the manual and the automatable parts. The decision was therefore taken to envisage a different type of planning, focussed rather on medium-term developments of products and manufacturing methods, and more in keeping with the main economic trends in the industry. Training with this objective is presently underway, but it can only be a stepping stone. *It is to a continual redistribution of expertise between the production planners, the workshop and the methods engineer, that TOTEM will owe its survival and its relevance.* Conversely, the latter will be a positive indication of the presence of this balance. Compartmentalization between the different actors, if it is too rigid, will rapidly sign the death warrant of the common knowledge base.

The use of TOTEM during the past few months has already demonstrated the importance of a new type of planning in the long-term, and such planning is now openly envisaged. The significance of these developments is clear: TOTEM is drawing the workshops in which it is installed towards more intense industrialization, at least in certain sectors. But the expert system itself was only the means of a first rupture; it was the project that was the real driving force of evolution since it mobilized, over a period of several years, a high level engineer around the most commonplace practices of the workshop. This exceptional intellectual investment is, beyond TOTEM, the fundamental vector of a changing industrial paradigm.

Nevertheless, TOTEM is not yet the pivot of a set of production management systems. Was the initial perception wrong, or over-ambitious? One can hardly answer such a question in the absolute, but it seems rather as though the emphasis has shifted towards production modes themselves, owing partly to TOTEM. *Learning from a project is also learning what the limits of the project are.*

3.2 GESPI: from station traffic planner to network planner

Out of all the projects studied, GESPI was the one that took the most time to be operational and required the most significant changes to its original design. The complexity of the station planners' work, the constraints they had to deal with daily and the surprises revealed in the study of the station's infrastructures or of the movements of trains, all contributed to several renewed attempts at developing a system, different to the original plan. The first operational tests also proved to be a

3 Dynamics of the projects: multiple lines of transformation

severe trial for the project. Questions arose as to whether the system should be left to make certain tricky decisions, whether the station planners could transgress certain constraints that they had themselves imposed, or what to do when the system rejected apparently acceptable solutions. It was thanks to the direct involvement of its designers during the entire implementation phase, that the system's survival gradually became a reality. But can it be said that GESPI, like TOTEM, promoted a new kind of engineering?

This does not seem to have been the case, even though the maintenance of the software could not be left to the planning team alone. In GESPI there was no reconstruction of what could be a "station entry" theory, comparable to the elements of a rolling or wire drawing theory for TOTEM. The system remained too close to the planners' own ad hoc decisions, and its structure was too sensitive to certain adjustments to the infrastructure or train schedules. However, by representing the station planners' constraints in concrete terms and by relying on the increased computerization of all GESPI's peripheral functions, a shift in the position of the planning service was started. The project brought to light the pressure and multiple incidents that the planning service has to absorb as efficiently as possible. But does the planning of maintenance work on the tracks, or changes to the schedule or equipment, take the station's limits into account? This may seem to be the case, although GESPI shows precisely that the formulation and expression of these limits are by no means easy.

Thus, GESPI can increase the station planners' argumentative capacity, either by making preliminary simulation – and thereby different choices – possible, or by legitimizing their expertise and their detailed knowledge of the station. The development of GESPI showed that the planners were not just users of ready-made recipes and that they were capable of evaluating the risks associated with certain operational configurations. *GESPI therefore tended to institute the planning service as a network manager, and not just as an office for drafting plans.* This amounted to finally acknowledging the evolution of the station where the expansion and diversification of its activities over the past ten years have progressively given station planning a mission. GESPI has helped to reveal the difficulty and the true nature of that mission. In this sense, it has participated in the rehabilitation of the operator's expertise, a trend that is also perceptible in other industries[6].

3.3 Cornélius: the problem of transferring expertise

TOTEM and GESPI restructured experts' know-how, yet remained the instruments of those whose knowledge they embodied; they were systems for experts.

---

6 Midler C. (1988); Fixari D., Hatchuel A. (1990).

This was not the objective in the Cornélius project. It aimed at transferring the knowledge of a maintenance specialist responsible for fine-tuning a machining flexible cell that was essential in the production process of a workshop.

The group responsible for designing the project did not encounter any particular difficulties at first. The expert had no problems in adapting to the structure of the software, especially since he had already tried to systemize his knowledge in the form of manuals and recommendations. The work with Cornélius was fruitful and helped to increase his knowledge and stimulate thought on strategies for making the facility more reliable. But was Cornélius transferable, and if so, to whom? Two actors were possible candidates: the local workshop maintenance fitters, or the operators themselves. At that stage uncertainty concerning the nature of the user weighed on the project. It quickly became clear that the modelling opted for was too closely adapted to the expert's concepts and particularly to his capacities for intervention, and that it necessitated certain investigations or an understanding of certain facts.

Thus, the order in which the system was to ask the questions that guided the diagnosis appeared too "logical" for operators used to reacting in a systematic way without completing an analysis of the causes of failure. Moreover, the tests to be carried out to answer the questions were beyond the operators, or included certain ambiguities which could not be cleared without the help of the expert. The difficulties in transferring Cornélius were not signs of an insurmountable barrier between the expert, the local maintenance fitters, and the operators. They showed rather that the expert's know-how was not directly transferable and that a simultaneous restructuring of this know-how and that of the operators was necessary to obtain a tool which was useful to the latter. This would mean developing a new type of operator with extended knowledge and capabilities. Could Cornélius alone warrant such a change? Perhaps, but at the same time the stakes associated with the project diminished. The machine became more reliable, thanks partly to the operators who had constituted their own expertise on the job, and partly to the initial work carried out by the expert with the help of the system.

Could the Cornélius project have had a different history? We do not think so. The first stage, which consisted of modelling the maintenance expert's knowledge, was essential. Yet by intensifying a critical analysis of the facility, it initiated a race between increasing the facility's reliability and, if this was not successful, transferring broader maintenance expertise to the operators. It was by allowing for a better exploration of these two alternatives that Cornélius revealed the issues at stake. The transfer of maintenance expertise foreseen as the project's initial objective, was evidently less urgent than the production of knowledge to increase the facility's reliability.

The Cornélius experience illustrates clearly this dialectic between the sharing of existing expertise enabling everyone to act effectively and the dynamics of the production of expertise which can make this sharing less useful; for in order to preserve knowledge, it must at least have sense.

Even though it takes place in a very different context, Naval demonstrates this same observation in a different form.

3.4 Naval: the loss of actors and their expertise

As we have just seen, Cornélius grew from a technical rupture. The new flexible cell required special training as well as the creation of strategies for increasing its reliability. But the production of knowledge can be the result of crises of a totally different nature. The Naval project resulted from an abrupt change in an oil company's economic environment, resulting in an organizational crisis. These ruptures which led to the birth of Naval were also the cause of its discontinuation when the reverse occured.

The oil shocks of 1974 and 1979 severly shook the hydrocarbon market. The latter gave rise to two years of euphoria and hence to a boom of exploration activities. The oilrig market was the scene of extreme tension and the oil companies had to accept the constraints of renting rigs at exorbitant rates and for long periods (up to three or four years). We saw in preceding chapters the problems that these companies encountered from the beginning of the 1980s, when the price of crude oil stabilized; they had to try and use rented rigs as far as possible and to avoid any new commitments.

Naval was conceived as a tool to support this strategy. It was expected to produce a schedule which complied with the technical constraints involved in the use of rigs, and which was also acceptable to the different actors in the company directly concerned in exploration – subsidiaries, headoffice, and technicians. However, the expert system was unable to capture the knowledge of the experts officially assigned this planning job, since the job as such had never existed. The project was therefore, in itself, the medium conferring the title of "expert" on certain actors, who were moreover the only ones to "carry" Naval and give it its contents. But this legitimacy, in turn, affected the company's organizational logics. If Naval was a recognized tool, it was henceforth possible for certain actors to thoroughly prepare, even to "steer" negotiations on the utilization of oilrigs.

Between 1982 and 1985, these negotiations were subjected to increasingly specific protocol. Central and regional bodies were set up, and rules limiting the subsidiaries' autonomy were decreed; in particular they required head office approval for any rental contract exceeding six months. But the technical nature of each case, the heterogeneity of each subsidiary's constraints, and the variability of evaluations concerning the feasibility of exploration programmes, made the negotiations difficult and sometimes frustrating for head office. On several occasions the executive responsible for leading the negotiations expressed his reservations as to the quality of available information. It was, however, uncertain whether he would be able to obtain the partners' agreement on the use of Naval as a filter, a type of safety barrier to control the results of the negotiations. The question remained open throughout the duration of the project. The two experts, who were also responsible for the project, considered that the nature of their role in the company's senior management would be shown in the answer to this question. These actors, who had until then been responsible for, respectively, the company's secretariat and strategic co-ordination, and the advisory services to regional subsidiaries, were in favour of strengthening the negotiation procedures concerning oilrigs. They felt that Naval could be the ideal instrument in such a strategy. It required the rigorous maintenance of case files concerning each prospect and an indication by each partner of the values attributed to each planning parameter. Naval could also favour the development, in each of the subsidiaries, of investigations prior to the options proposed, according to a common methodology.

The counter-shock in 1986 considerably limited the relevance of a strategy of strengthening negotiation procedures. The decline in crude oil prices was followed by the collapse of the oilrig market and the drastic reduction of exploration programmes. Oil companies were then able to lease rigs as the need arose, and for the required period. The co-ordination between subsidiaries was no longer useful, bar a few exceptions.

What was the sense then of such a project? *A priori* Naval was not, for all that, any less relevant. One could even argue that if another crisis occured, it would be able to provide a means of preventive control to avoid the proliferation of incoherent decisions like in 1979-80, or help to identify the moments when the company could take the risk of signing long-term contracts at low rates. This strategy was, however, to prove unfeasible. The company's restructuring in the wake of the second oil shock directly affected the group responsible for the project since the two main experts went on early retirement. Together with them, experience of the 1975, 1979, 1982 and 1986 crises disappeared. A technical and social asset was lost, for their experience was based as much on their knowledge of drilling as on the network of relationships which made them acceptable mediators to the different negotiating parties.

Naval is from this point of view a very good example. *It is rare to see with such clarity how an "actor" is formed by a specific combination of expertise and relationships, and how the actor disappears when this combination is considered obsolete.*

To which model of change do such examples correspond? We show that to understand them better one has to define certain notions more clearly and to look beyond the most classical perceptions of organizations. This is due, in particular, to the fact that most of those who have taken an interest in these subjects have not paid attention to the production of expertise in organizations.

4 Organizational change: production of expertise and metamorphosis of actors

Technological change in business and industry has given rise to a profusion of analyses and interpretations over the past fifty years. In the 1950s, several studies tried to shed light on the social repercussions of new types of work introduced by mechanization and automation. They looked at workers' new autonomy and qualifications, or at the problems that traditional foremen faced when trying to maintain their authority in these new technological systems[7]. But most of this research dealt with new technological systems that had already been set up, and consisted of analysing the consequences of this transformation *a posteriori*. The process of change was therefore studied only in the light of what emerged once it was complete. This type of analysis is suited to changes following a decisional process on different hierarchical levels, during which a design department first determines the structure of a workshop and its organization and then leaves it up to the operational teams to adapt to the new context. At the start of the 1980s, this type of planning was severely criticized, in favour of an approach which considered both technological systems and the kinds of job or behavioural pattern they were likely to promote. The design of "reskilling" organizations[8] and the taking into account of downgraded functioning and job enrichment, were then amongst the new objectives assigned to planning. Moreover, such objectives did not seem realistic without the involvement in the design process of the operators themselves, wherever possible. Today, several experiments have been undertaken[9] with the purpose of effectively, even if it is only partially, associating the definition of technological systems with their future users.

Such practices have significant methodological implications for the researcher, for to understand the effects of these changes it is necessary to reconstitute the different actors' trajectories and to follow both the way in which expertise is exchanged and the course of the project itself. *The involvement of operators in planning is often associated with the idea of "negotiated modernization". The notion is not intended to raise theoretical debate but, by emphasizing the idea of negotiation, the process of change is given an exclusively "political" image, as if the issues involved were limited to the opposition of interests between well defined actors.* This concept can also imply that each party knows what it wants to negotiate and that coherent solutions or well-defined alternatives are determined and defended by the partners in the negotiation. In this model, the arguments used do not

---

7 For an excellent synthesis of this question see Michael Rose (1988).
8 Riboud A. (1987).
9 Midler C. (1988); Walton R.E., Susman G.I. (1987); Whittaker D.H. (1990).

reflect the expression of each party's expertise as much as a rhetoric used chiefly to support predetermined positions.

The model has its relevance in many situations, but does not lend itself very well to the process of change, where solutions cannot be found immediately and have to be constructed progressively. In other words, it is ill-suited to situations in which the desired collective functioning cannot be accurately defined, and where there is room for an open-ended process of collective learning and innovation[10]. This is a process which characterizes more and more industrial situations, such as project groups in which several services participate, or organizations in which certain persons have to simultaneously define the contents of a mission and the conditions of its implementation. Projects for the development of expert systems or, more generally, rationalization projects can belong to this type of process. *One of its essential features, which we consider important enough to warrant closer attention, lies in what can be called "a metamorphosis of actors". We encountered several examples of it in the projects studied.*

Use of the notion of metamorphosis serves a specific theoretical objective. This consists of emphasizing that to consider real changes as mere adaptations by the actors at the start of the project, or as the result of a shift of these actors' control or ability to intervene, is an inaccurate interpretation. In fact quite the opposite takes place, *since it is not the actors whose behaviour changes, but rather the transformation of these actors which leads to new behavioural patterns.* For, although the individuals concerned remain the same, it is their way of thinking about themselves and of recognizing themselves as "actors" in a particular group, that may change. Their position, the sense of their action, and the conditions of access to the jobs they occupy are all transformed – without this excluding *the birth of new actors* (a borderline case of metamorphosis) i.e. the definition and acceptance of positions and of expertise until then non-existent in the organization[11]. The birth of an actor does not necessarily mean that the individual embodying it is new in an organization, but rather that it was possible to invent a new figure that is viable and relevant.

The notion of actors' metamorphosis has two advantages: it shows the limits of more traditional perceptions of organizations and makes it necessary to define the concept of an actor more clearly.

---

10 Hatchuel A. (1988b); Charue F. (1991); Senge P., Morris L. (1993); Midler C. (1993).
11 It is, of course, possible to consider any metamorphosis as a kind of birth, but we prefer maintaining a difference of degree between these two notions. If the project results, for example, in a new way of being a production planner or a station planner, this is closer to metamorphosis, whereas the creation of a maintenance technician where one did not exist before, is obviously a birth.

4.1 The limits of organization seen as a game

The purpose of an image or a metaphor lies in the phenomena that it highlights and to which it draws attention. There is an old tradition which tends to interpret collective or organizational behaviour by means of a particular image: that of a game[12]. *This image is particularly relevent when it aims at highlighting individuals' strategies in a formal framework known to most actors.* At the beginning of the century already Taylor described how piecework systems led employers to try continually to decrease the unit price, and symmetrically, how workers tended to slow down collectively to prevent them from doing so. Such behavioural mechanisms obviously have universal traits: as soon as a rule, a contract or a convention defines the interests, prerogatives or duties of the contracting party, it can be expected that the individuals concerned will use to their own advantage all the leeway they have. The expression "getting round the law" translates this tendency very well and although there is nothing forcing anybody to do so, it implies that whoever does, has sound knowledge of the law.

In this type of model, change can first be considered as a modification of negotiating positions within a well known game. If a particular individual is a better actor, and therefore strategist, than the others, the benefits are shifted to his or her advantage. A more complex view is possible by imagining a modification of the law itself, thus of the rules of the game. The constitutions of States formulate systems of rules which stipulate when and how the law can be changed, but in the final analysis they remain no more than rules for the end of the game, since no convention or organization is able to define in advance all the possible forms it could take. *We thus reach the end of the game metaphor when we want to understand how a process of change is started without being able to predetermine the rules of the final game it might end up to be.* For what suddenly appears and goes beyond the initial game, is the production and recognition of new knowledge leading to a different interpretation of the nature and meaning of the group itself. *The notion of actors' metamorphosis is therefore intimately linked to the emergence of new knowledge and to its diffusion amongst the individuals concerned.* It is this new expertise which makes the establishment of new relationships, and consequently the acceptability of new actors, possible.

4.1.1 Individuals and actor figures

To define the idea of the metamorphosis of actors even more clearly, it must be noted that an organization is more than a mere collection of individuals and more

---

12 Crozier M., Friedberg E. (1980).

than a group of contracts linking individuals to one another[13]. It necessitates both the division of tasks and the search for suitable forms of co-operation in the face of the multitude of situations which make up the daily life of a firm. The production planner, the station planner and the maintenance technician all act, invent, negotiate and simultaneously explore the acceptable meaning of the notion of production planner, station planner, or maintenance technician. It is therefore necessary for these notions to be developed collectively, and to denote what can be called "figures of acceptable actors". Whilst the word actor refers to the person or group of persons that follow a specific action logic, the term "actor figure" makes it possible to denote more specifically the set of conceptions, expertise and procedures which allow this action logic to be recognizable and acceptable. Stated more simply, it amounts to distinguishing clearly between a specific doctor (the actor) and everything that makes the notion of physician (actor figure) acceptable and operational.

When talking of the metamorphosis of actors, we are specifically interested in the transformation of figures. The methods engineer born with TOTEM or a new conception of station planners related to GESPI are such figures, whose viability cannot be predicted. Each person can of course modify an actor figure by his or her talent and personality, but the modification has to be acceptable to others. *An actor figure, in the sense we have given to the term, is not only an institution, a functional structure or a set of rules; it is a set of potential intervention recognized as legitimate, valid and acceptable in the framework of an organization.* If any one of these criteria is challenged, the figure has to evolve or disappear. Thus, the Naval experts had a lot of difficulty in gaining acceptance for the new figure of a planner in the complex context of the company's exploration activities, but they themselves were threatened by the oil counter-shock.

When we refer to an actor figure that history has long associated with recognizable competence and modes of intervention, we talk of a "trade" or "profession". But that is an extreme case. The creation of new actors – and therefore the stabilization of new figures – in firms cannot always have the clear and lasting character normally associated with the idea of a trade. It is a more complex process which depends, as we have seen, both on the production of new expertise and on the correlative implementation of new networks of relationships.

Reorganization and restructuring are fashionable words in corporate language, but their commonness masks the real stakes involved; it is not that easy to create valid, legitimate and acceptable actor figures. Furthermore, the latter also have to depend on a new distribution of expertise and be able to build a network of viable relationships. The danger of reorganization not born from the true shift of exper-

---

13 Recent trends in economics use this representation of organizations. It is adequate for an interpretation of certain phenomena in the present evolution of firms, but too restrictive for the analysis of dynamics studied here.

tise, or which cannot allow the latter to appear because everything is supposed to be known in advance, lies here and nowhere else.

4.1.2 Actor figures: identity logic and societal logic

Referring to the metamorphosis of actors is using an image again, and analysing its basis and theoretical limits would require an in-depth discussion. One should not forget that such metamorphoses can conflict with identity or institutional logics. In the projects that we studied, these logics did not play a determining role in the transformations observed, but we could not exclude the possibility and, even if our research perspectives had been different, these problems had to be borne in mind. It is fairly easy to imagine that the creation or transformation of an actor figure can clash with strong misgivings in the persons attached to former ways. Several sociological studies devoted to these questions[14] have explored the way in which individuals experience certain work situations and actor figures as veritable identities. For them, leaving these situations does not only mean changing their habits, it means changing their socio-professional references completely. Thus, certain changes in processes inducing the metamorphosis of actors are far more profound than a loss of power or the restriction of individuals' scope of intervention.

The authors of a recent study[15] noted the development in certain people of *"corporate identities, replacing rigid professional identities"*. The individuals concerned usually have a large capacity to follow in-service training likely to lead them to responsibilities which differ greatly from those that they held initially. Do such identities also signify a greater ability to cope with the chaotic nature and uncertainty of projects like the ones that we studied? Our research does not allow us to draw such a conclusion. Furthermore, the transformations which we observed in the projects were not radical enough for the "professional identities" of certain persons to be challenged. Thus, the metamorphosis of actors that we describe has a continuous aspect, and this continuity is facilitated by expertise which can overlap or is easily exchanged without the actors being completely versatile. In this case one can understand that organizational functioning was modified by the appearance of new expertise, without the individuals concerned losing their identities.

Since a firm is not "an empire within an empire", the creation of new actor figures can also clash with the rigidity of systems of classification or definition of salaries, with the prevailing rules defining trades, or with existing links with the education system (i.e. qualifications required). This *"societal"* effect has been

---

14 Sainsaulieu R. (1985).
15 Dubar C. (1991).

analyzed[16] very accurately by means of international comparisons, and shows that the transformation which takes place may have to play with the rules. The effect is not, however, one way for it seems that the processes of metamorphosis of actors could be intense enough to lead to a revision of institutional logics, if these prove to be too rigid. There have indeed been cases of renegotiation of classifications recently. Finally, the metamorphosis of actors does not necessarily create new well-identified occupations; it therefore seems that the emergence of stabilized occupations is merely a particular case in the processes of corporate transformation.

The projects that we studied mostly concerned a small number of persons so that institutional problems such as these hardly influenced the course of events, or were only implicit. In any event, even if "societal" effects do control and shape the process of change, they are not its source. Moreover, even if the metamorphosis of actors cannot be totally free of multiple institutional and sociological regulations, it is still necessary to identify accurately the conditions in which it emerges.

4.2 Production of expertise and control of the framework of collective action

We can therefore understand why, as soon as there is a possibility of new expertise being produced, the process underway cannot be interpreted exclusively by a view of corporate life that is too "political". This type of view, which has become increasingly widespread since the sixties[17], gives prime importance to power relationships in the interpretation of organizational behaviour and change. Its emphasis on the study of power positions, of actors' latitude, and consequently of all the games which may result, is indeed useful. However, such an analysis, carried out at the start of the project, would depend implicitly on a state of expertise and on its distribution between the persons concerned, which the project's history will then disrupt. It is of course possible, at some stage during the project, to locate the "zones of uncertainty" that each actor will try to defend or control, but one soon realizes that the contents of these notions fluctuate constantly for the actors themselves. *They no longer appear as stable frameworks which make action intelligible, but as the transitory effects of a form of regulation at work in the permanent opposition of expertise underlying relationships formed during the project.*

It is, however, still relevent to analyse how this opposition of expertise is itself influenced by institutional positions or by the systems in place. Thus, throughout the process, the actors work on the "truth" of the project, on its "progress", on

---

16   Maurice M. et al. (1986).
17   Crozier M., Friedberg E. (1980); Pfeffer J. (1981); Mintzberg H. (1983).

4 Organizational change: production of expertise and metamorphosis of actors

its "possible results", on its "efficiency", on its "future implications" and on the way in which the initial actor figures might be transformed. In turn this regulation influences future decisions.

Out of the four projects studied, Naval was the subject of a particular experience. We were involved in the project as participant observers, trying to intervene at certain key moments, to reconstruct the opposing arguments and analyse their evolution and mode of validation[18]. We were thereby better able to appreciate how the opposition of expertise plays two distinct roles: one as an outcome of the project itself, the other as a regulator of its management.

Analyses which assume that experts who reveal their expertise lose their influence and power, are therefore far too simplistic. They see expertise as a weapon that can be passed from hand to hand. Of course a manufacturing secret passed from one expert to another in the same field can circulate in this way, but when it is a matter of comparing different expertise, the exchange is not so mechanical. It is sometimes by better explaining the conditions and contents of his expertise that an actor can highlight the complexity and importance of his work -as in the case of the production planners and station planners – on condition, of course, that these explanations are convincing.

This finally brings us to the conclusion that analyses adopting an exclusively "political" point of view, or one of negotiation, are insufficient to account for the dynamics and the viability of the processes of transformation. *They have created a tendency to neglect the dynamics of expertise as an effective regulator that shifts the references of action and opens the door to the process of metamorphosis of actors.* Since the process of rationalization implies, in principle, such dynamics of knowledge, we foresee that it will be effective only if it is accompanied by a new production of expertise and includes an acceptable metamorphosis of actors.

It is on these two conditions only that one can truly speak of "organizational learning". This concept appeared at the end of the seventies[19], but is now the object of renewed interest that is warranted, considering the intensity of the transformation and rationalization processes on which all firms have embarked. However, organizational learning is not a natural attribute of organizations, it is the consequence of the kind of project management which understands the dynamics of knowledge and remains aware of its effects on the actor figures involved.

On this basis, the main issues in expert system projects are easily defined. But it is possible to go even further, and, by taking a historical standpoint, highlight the true nature of management techniques.

---

18 This aspect of the project is discussed more thoroughly in Part Two of the book. See the chapter "Naval – Undefinable expertise of strategic planners".
19 See Argyris C., Schön D.A. (1978).

Chapter 4
The nature of management techniques
Dynamics and unexpected repercussions of rationalization

What general conclusion can be drawn from our four case studies and which of its main features would be specifically relevant to the approach and contents of expert systems? In the preceding chapters we tried to gain insight into the issues involved in expert systems by adopting two viewpoints simultaneously. One can be compared to that of a technician or engineer, since it consisted of revealing the basic hypotheses of expert systems which are too limited for the diversity of expertise in practice; the other can be qualified as organizational or sociological because it studied the dynamics of actors involved in the project.

The history of an expert system cannot be understood only in terms of the problems encountered with respect to modelling, it is also closely linked to the actors, their challenges and, as we have seen, their metamorphoses. These two points of view do not correspond to two distinctly separate paradigms, nor to a mere sociotechnical juxtaposition where each one could be interpreted alone. *The two types of phenomenon interact constantly, which means that any attempt to comprehend one implicitly calls for hypotheses on the other*. It is therefore understandable that organizational phenomena cannot be reduced to either a pure engineering science or to pure sociology. The failure to acknowledge this fact – either collectively or scientifically – amply explains why two centuries after the birth of industrial enterprise there is still a rift between the two viewpoints on both the conceptual and practical levels.

In particular, the consequences of this type of rift hinder an understanding of organizational processes in which intense interaction exists between the two dimensions. This is precisely the case in any *rationalization* project, because it institutes a technical development plan and because it may well involve the formation of new actor figures. Thus, each wave of rationalization has given rise to a chain of misunderstandings which has characterized most management techniques precisely because their implementation was necessarily a rationalization project. *Our attempt to define the nature of management techniques therefore brings us closer to the very essence of the organizational phenomenon, to that which makes it neither a general structure of abstract rules, nor a mere cultural symbiosis between several individuals.*

This chapter is devoted to that attempt. We first highlight the main features of the projects studied, which allows us to put expert systems into perspective vis-à-vis the main waves of rationalization that have punctuated industrial life during the past century. We then show that all types of rationalization have a common, incomplete, structure.

1 Portrait of an expert system project: from myth to stakes

Summarizing the history of this type of project obviously does not amount to outlining a rough plan with three or four main chronological stages. It is far more useful to discuss the main issues which inevitably arise and whose conclusions or solutions largely determine the project's physionomy. There are three such issues: *the constitution of a mobilizing myth, the reformulation of expertise, and the development of stakes.*

1.1 A mobilizing myth

Expert system projects are the doing of actors who embark on this adventure with a somewhat mythologized description of an industrial problem. Their limited knowledge of the project's industrial context at the time it is launched is a feature that is all the more remarkable for being common in all cases. Moreover, these actors' interests or objectives differ considerably from one project to another. Whether they essentially want to promote artificial intelligence, develop a new production strategy or simply alleviate certain thorny problems, makes no difference to the fact that the project is mainly, for them, a trip into a microcosm of questions and practices with which they are largely unfamiliar.

This dimension is inevitably present in all innovation, for how can one avoid a minimal dose of utopia when embarking on the unknown? However, the idea needs to be qualified since the contents of this utopia are not the same as those found in the idea of sending Man into outer space. An expert system is *a priori* feasible with existing technological means and none of the projects studied necessitated a truly new technology, even if certain technical problems were encountered every time.

The utopia lies elsewhere, in the idea that the project will lead to acceptable economic and organizational progress. It is in relation to this perspective that the mythologization of industrial problematics constitutes both a driving force and a

1 Portrait of an expert system project: from myth to stakes   87

risk for the project. It may seem relevant to ask whether the situation could not be analysed more thoroughly before the project was launched? The answer is that some technical or economic scenarios could, certainly, be worked out and a progressive step by step approach adopted, but at some stage one always has to "pay to see".

1.2 Enhancing and sharing knowledge

The enhancement and sharing of knowledge is certainly a direct result of what can be called the modest ambition of artificial intelligence. The initial phases of all the projects studied, which lasted far longer than anticipated, were mainly devoted to a detailed study of the information provided by the selected experts. The successive to and fro movement during the modelling phase, punctuated by the need to have the experts validate the results obtained, provided continual stimulation for the exploration of a problem and its context. Project leaders spent many long months working with individuals who were often on a very different hierarchical level to their own, since station planners, production planners and maintenance technicians served as experts. In contrast, in the Naval case two young AI specialists had to question seniour oil company managers in detail, and sometimes even challenge the coherence of their expertise.

This interchange was in itself an organizational modality that differed totally from the daily functioning of the firms in which the projects took place. *An essential element in the life of a project lies here, for the project would be totally impossible if the actors all kept to their normal place in the organization.* All innovation implies that certain actors can, in some way or another, intervene in the activities of others. Those recognized as leaders or organizers naturally have little difficulty in that respect. In the expert system projects studied all the work was carried out with managerial approval, but the experience did not lend itself to a model of asserted authority between the different actors in the project and particularly not between project designers and experts; it was the essential aim of modelling knowledge that prevailed. There was, moreover, no case of resistance from the experts or accusations by them of interference, although it should be noted that none of the projects presented a real threat to them or their jobs (even though some staff transfers did result from related productivity gains). When at the conclusion of the Naval project the experts benefited from an early retirement plan, this was due to a harsh economic crisis in no way related to the impact of the project itself.

The salient feature of all these projects was therefore that of a forward movement involving actors whose know-how was solicited and who could take advantage of the opportunity to bring to light the constraints with which they had to cope and even the shortcomings of their action. The main result was a broadening of knowledge shared by the actors, with the inevitable destabilizing and unexpected impact of this type of phenomenon.

1.3 Transformation of expertise and discovery of stakes

The modelling carried out to achieve the desired level of automation is not, however, a simple matter of the transfer of know-how from the expert to the system designer. The designer has to transform the knowledge, which sometimes means expanding it by incorporating new knowledge. In this sense, it would be a mistake to see the modelling process as nothing but the capturing of experts' knowledge, for it necessarily implies the choice and reformulation of knowledge. There is some knowledge that can, some that cannot, and some that does not warrant being automated. It is in relation to problems such as these that the different actors will define the stakes they see in a project. These may be based on a number of criteria, such as complexity of the software, feasibility of knowledge, maintenance capacities, nature and status of users, and so forth. Thus, the stakes in the project cannot be known until the transformation of knowledge, implied in automation, has been defined. Only at that point can the exact scope of the software be appreciated, while the potential relationships that it promotes or undermines are taking shape.

Not talking about stakes until the last stages of a project is an apparent paradox. However, if one bears in mind that the project is born of the mythologization of an industrial problem, the extent of its reach can obviously not be determined until the true potential of the expert system is better understood. The project's survival therefore depends largely on the discovery of an automation strategy corresponding to issues which are important enough to cause the shift of a network of relations or the emergence of new actors capable of keeping the system alive. Without these dynamics it will rapidly disappear, *although the project's impact will not be lost entirely. The effort at sharing and transforming knowledge has its own consequences which can promote new and more efficient strategies, even if the advantages of automation are challenged.*

There is something very specific, yet also very general, in these problematics. Even if a project has its originality, it also has something very familiar about it. Devoting intelligence and time to certain expertise means following the same old path of rationalization, with its inevitable mirages and its acceptable oases. But this consists less of exploring a new area every time, than of taking a different view of familiar ground. By trying to situate expert systems in a broader perspective and by viewing them as a medium for a new wave of rationalization, we can understand the characteristics of these waves more easily and see beyond the systems themselves.

We shall now undertake a first comparison by defining the similarities and differences between expert systems on the one hand, and operational research on the other. We shall then propose a more general conception of management techniques *which will help to explain why each of these historical experiments, by way of the questions it spawned, revealed an unrecognized aspect of collective action. It was these successive revelations which probably constituted an essential component of the productivity of management techniques.*

2 From operational research to expert systems: a new industrial logic

From the outset, expert systems and more generally artificial intelligence, gave rise to debate and controversy. The appearance of research labeled cognitive science and headed by slogans such as "The intelligence revolution", rapidly moved the debate away from industrial realities towards a more theoretical questioning. Although the latter did have its philosophical worth, it also helped to bathe AI in a metaphysical halo that was criticized by certain specialists for placing AI in a negative light vis-à-vis the business world[1]. A few years ago certain authors[2] also wondered whether "expert systems were not a programming technique like any other". The question has, moreover, become all the more relevant since it was later revealed that many projects initially written in languages believed to be characteristic of AI[3] and which were in a sense its signature, were rewritten afterwards in conventional computer languages[4]. Are expert systems not then simply another way of writing programs?

Yes they are, but the observation does not mean much. One could rather wittily argue that in computing, like in literature, "it's the style that counts". The important thing, no matter what language is used, is always the modelling process which turns the computer software into a tool. Expert systems have an objective which is not shared by all computerization and which is situated in the shift from the notion of information to that of systems of knowledge or expertise; in other words, not a mere description of objects and relations between these objects, but a set of propositions organized in dynamic terms of causality, goal attainment, evaluation,

---

1 *01 Informatique*, 29 May 1989, "Dossier IA": *certaines firmes déclarent préférer faire de l'intelligence artificielle sans en parler* (certain firms prefer using AI without making this known).
2 Marois T. (1985).
3 Lisp or Prolog, in particular.
4 Chauvet J.M. (1987).

and so forth. One could argue that any computer program can be treated similarly, yet, even though that is true, it simply highlights the fact that a single problem can be modelled in several different ways, depending on its context. The argument is also put forward that AI effectively exists only when the system can learn to modify its reactions and rules. That is acceptable, yet the process can be simplified to the extreme. If industrial experiences are anything to go by, the prevailing view of expert systems is still the classical one.

The specificity of expert systems can, nevertheless, be contested more legitimately by an older discipline. The modelling process used by expert systems has something in common with operational research, a discipline which has now slipped into the background but which was, until the end of the seventies, almost as renowned as expert systems are today. A comparison between the two techiques, or projects, is therefore particularly useful to identify both their common points and their differences.

2.1 Operational research: modelling to optimize

The basic pattern of an operational research project is the following: in order to resolve a particular problem, a model is developed which summarizes available knowledge (we have used AI language intentionally), generates possible answers to the initial problem and, where relevant, makes it possible to select the best one. Amongst the most classical and still relevant examples of the approach is the very widespread case of monthly planning in oil refineries[5]. Schedules are prepared by means of a computer program which describes, in the form of linear equations, the main rules for the functioning of the refinery. The software also looks for the most suitable procedures for the different industrial units comprising the refinery. In particular, it takes into account in its "reasoning" the prices of the different products involved as well as the quantities to be ordered and delivered, by applying an economic function that it tries to maximise – in general the profits obtained by operating the refinery in relation to a particular market. This example is well-known mainly because the calculation makes use of a famous mathematical method which always gives rise to active research: linear programming. But the utilization of this type of technique does not in itself typify an operational research project. The latter can be identified far better in the following general pattern:

- The actors concerned are on the one hand "deciders" and, on the other hand, "operational researchers".
- The operational researcher gives the decider the means to choose the "best" solution to his or her problem, or at least a solution that is more rigorously and

---

5 Renouard O. (1988).

more scientifically validated than the one he or she normally uses to deal with the problem concerned.[6]

After WWII there was a progressive broadening of what was understood by "the best solution"[7], *but the essence of operational research remained linked to the idea of enhanced efficiency provided by the operational researcher for the decider.* This clearly meant that the computer systems developed had to do better than human experts because they used different methods of reasoning or calculation. In the sixties and seventies, the idea that an operational research model could produce results equivalent to those of a human expert virtually disqualified the technique because of the costs involved. The demand for the superiority of the model diminished slightly with the so-called "decision support" movement. The aim was no longer guaranteed optimization, but the idea of increased coherence and relevance in the solutions proposed was retained. Although very brief, this description of operational research shows how it both resembled and differed from artificial intelligence or expert systems; while OR and AI had a similar purpose, they did not have the same objectives.

The idea of *a similar purpose* is somewhat obvious. Both projects were designed to formalize or model an intellectual activity dedicated to the resolution of a specific problem. This similarity is in no way diminished by the fact that expert systems use programming languages or software designs unknown to operational researchers, nor by the fact that expert systems are far more user-friendly than traditional operational research programs.

As far as *conflicting objectives* are concerned, the specific aim of operational research emphasizes the researcher's contribution to the decider, whereas expert sytems place the cognitician in a position of listening and receiving vis-à-vis the person designated as the "expert". Moreover, the terms used effectively translate the inversion of the philosophies, at least in their most common form. *The operational researcher who knows how to solve a problem contrasts with the image of a cognitician who merely captures others' knowledge.* The decider looking for help or at least for a clue to find the best solutions to his difficulties, contrasts with the expert who possesses expertise of which he is sometimes only vaguely aware, but who is considered as a source from which knowledge need only be drawn. The reasons for these differences lie in the history of each of the two disciplines and therefore in the context in which they were conceived.

---

6 Moisdon J.C. (1985).
7 Roy B. (1990).

2.2 Operational research and expert systems: different origins

The concept of operational research was first formulated as an organizational response to the appearance of complex pluridisciplinary problems. Recent research shows that even before the most frequently quoted examples of logistic or transport problems, it was the impact of radar[8] on the utilization of air defence systems that was a major preoccupation and the reason for founding the first team labelled "operations research". Its work did, indeed, consist of research on "operations" in the military sense of the word, thus not on weapons themselves but on their use in the operations sphere. Faced with this exceptional technological change – the creation of a technique for long range detection – these first teams had to find and validate new methods of counter-attack. The use of mathematical tools was less significant here than the definition of the mission as being the invention of new ways of doing things. In other words, it was a mission of change and of progress. It was, moreover, only much later that operational research built up a mathematic arsenal which was progressively to be attributed specifically to it.

By comparison, the framework in which AI and expert systems were born was far more academic. A product of several successive research projects, the discipline remained a laboratory object for a long time[9]. From the outset it claimed to be fundamental research on reasoning and knowledge, which explains the particular emphasis on general aims such as imitating human reasoning or programming self-learning. It is thus understandable that the keyword of the discipline is unquestionably that of *knowledge representation*, an expression which necessarily evokes a descriptive attitude whereas operational research could be qualified as prescriptive. There was no idea of progress in the initial philosophy of expert systems, other than that resulting from automation. Whilst operational research was simply intended to be the application of scientific methods to practical questions, AI claimed to be a science with its own object: intelligence.

2.3 Expert systems: operational research dedicated to new industrial logics

A number of authors have already compared operational research to expert systems. Operational researchers consider ES to be no more than a modelling technique that is sometimes faster and more acceptable[10] than previous ones. They

---

8 Mac Closkey J.F. (1987).
9 Most treatises on the question give the following genealogy: first an initial project by Newell and Simon, which, after fruitless tests, led to simpler projects: games, pattern recognition and expert systems.
10 O'Keefe R.M. (1986).

2 From operational research to expert systems: a new industrial logic

therefore claim to reject any reference to notions of cognition or cogniticians, for which they see no sound basis, and believe that when it comes to knowledge modelling, their experience suffices. However, in articles on the subject in the American journal *Interfaces*, the dominant concern is for a more global and unified perception of what rather seems like interacting disciplines[11]. This view of a broader paradigm of operational research is also to be found in France, where expert systems are perceived as the natural extension of "heuristics", to which operational research resorted when it had no other method for finding the optimum solution. Expert systems are, however, expected to endow themselves with a more confirmed epistemology than that of operational research[12]. The reasons for such a favourable *a priori* attitude is the emphasis placed, by AI, on the project of modelling human intelligence, or on the impact that a better understanding of human intelligence might have on economic studies. Thus, the ties between operational research and expert systems are perceived differently, depending on the viewpoint adopted; pragmatic for those content to broaden the range of tools at their disposal, epistemological for those wanting to reveal the potential of a new area of science. These analyses all lack a much simpler observation: *what expert systems embody is the pursuit of the fundamental project of operational research, with renewed and wider industrial criteria of efficiency.*

Although the main purpose of modelling expertise is not to provide new answers to old questions, one cannot deny that its objectives assume a different, perhaps multiple, industrial meaning. There are several possible substitutes for the idea of optimization, so dear to conventional operational researchers. *Manipulable memorization, accelerated transmission, coherence control, rapidity of formulating an answer (even if it remains unchanged), facility of usage, transparency, and easy integration of several computer systems are all objectives which could be pursued by a project considering itself in keeping with the expert system philosophy.*

The significance attributed to criteria such as these unquestionably renews of the goals of operational research, for even when the latter used "heuristics" inspired by observed human behavioural patterns, it was more often to obtain better results than to meet one of the above objectives. It can therefore be said that expert systems appear as a new mode of industrial rationalization, because they are part of a modelling project formulated in a new way, rather than an autonomous discipline independent of all past forms of rationalization. Clearly, the relevance of such an approach is implicitly linked to particular industrial contexts and problems. In which cases is it better to retain existing answers and to try to automate them? When is it better to rather forget existing knowledge? This is a rough outline of the options which would characterize *a priori* the division between operational research and expert systems. Former experiences showed us, however, that

---

11 Fordyce K. et al. (1986).
12 Bourgine P., Le Moigne J.L. (1986).

in practice, the dividing line is often blurred; expert systems cannot escape from the need to "transform" existing know-how, or to incorporate knowledge other than that of experts.

Thus, if we tone down the differences between techniques used and relativize discourse on the imitation of human reasoning, expert systems appear in a different light, as a renewal of the older project of operational research, with the appearance of objectives and efficiency criteria which alone could not have justified such projects ten years ago. *It is effectively in the mutations of the industrial world that the significance of these criteria must be seen.*

2.4 Forgotten lessons of operational research: the problematics of integration

Placing expert systems in the perspective of older movements like operational research not only provides better insight into the issues involved in the projects themselves, it also teaches us a lot about the *simplification process* on which each of these movements leans. Thus, the comparison between the decider-researcher and expert-cognitician models not only highlights philosophical differences, it can also serve as a salutary reminder. Operational research continues to suffer from the ambiguous and far too limited nature of this conception; literature on OR wore itself out trying to explain the difficulties of the discipline by referring solely to the misfortunes of the decider/researcher couple. These were vain attempts, for three main reasons:

- The "decider" concept could not be posed by definition only, by concluding that if it could be defined, then it really did exist. Such a concept only had a chance of being operative only if it was related to a realistic conception of decision making processes in organizational life.
- The decider-researcher relationship, already caricatured by the theoretical model of the decider, was marked even more by a linear view of modelling: the decider states his or her problem and the researcher in turn provides a suitable model and solution. However, researchers found that the greatest difficulty did not always lie in *problem solving*, but rather in *problem finding*, in other words in identifying which problem was relevant. Consequently, modelling often led to an approach in which the relevant questions, expertise to be mobilized and objectives to be pursued, were brought to light. It also revealed the actors concerned who, active in existing decision making processes and suppliers of information or simply receivers of instructions, could influence the choice of solutions in their own way.
- Finally, operational research tended to convey as a postulate the idea that an organization spontaneously adopts any new solution as soon as its efficiency has

been proved. That was making the rather hasty assumption that it is always possible to evaluate all the consequences of a choice. This is unfortunately rarely the case and the uncertainty or variety of the effects of change generate both conflicts and reservations, as well as enthusiasm and interest. Operational research could not ignore these phenomena for a long time without equipping itself with a methodology for dealing with them effectively. In fact, several attempts were made to give operational research this framework which it needed so badly, but they did not achieve any real consensus and, more significantly, they never had enough time to do so. Operational research's image deteriorated very fast and, as a specialized department, has all but disappeared from firms. Without the spirit of operational research disappearing, it has taken on different forms such as autonomous mathematical disciplines, more contextual approaches and so forth.

The operational research story might have provided food for thought for expert systems, a warning that the ease of certain formulations does not mean that permanent reconstruction in the field can be neglected. But that would have necessitated a more general view expressing the common nature of all the waves of rationalization and thereby also of all management techniques.

3 On the nature of management techniques

The comparison which we have just made between artificial intelligence and operational research suggests the existence of a common structural project to which each approach contributes a different answer. But this remark can also be broadened to other management techniques. One can scarcely forget about scientific management at the start of the twentieth century[13] or not wonder about the similarity of these approaches to more recent movements such as of production management in the manufacturing industry[14]? Beyond the effects of mere fashion, these movements constituted, in their time, the main axes of change and characterized the principal stages of "rationalization". *Even if historians use the word to denote the period in which Taylorism was set up in the first decades of this century, there is in fact no reason to apply such a restriction. Every time that an industrial form appears more efficient and more viable in a given context, it is necessarily more rational and the efforts to establish it warrant the term rationalization.*

---

13   The reader is referred to the following for an approach to the problems of Taylorism: de Montmollin M., Pastre O. (1984); Moutet A. (1985); Rose M. (1988).
14   Hatchuel A., Sardas J.C. (1990).

Rationalization is a mythical objective, a figure of progress in firms to which each period, each main management technique, temporarily gives more substance – meaning the conceptual and practical means to make of it an action programme. All such figures have known similar life curves. In their infancy they were met with enthusiasm or reservations and fed by vague or mythical images describing the progress or calamities to be expected. These phenomena can be expected for any kind of technology (e.g. early debate on electricity or the automobile), but they are particularly intense in the domain of management techniques which appear more abstract, more complex and more difficult to evaluate or define, even though they display very tempting ambitions.

Scientific management claimed to replace labour conflict with "harmony and concord"[15]. For many, operational research embodied the hope of introducing more science into the running of economic affairs. Computer aided manufacturing, in turn, led to nothing less than a "crusade" by certain American associations, in the name of a new industrial ideal[16], while artificial intelligence continues to this day to feed hopes for thinking machines. This early enthusiasm is inevitably followed, after a few years and once the practical problems have been revealed, by certain reservations or even outright disillusionment. At the same time, means of ensuring survival in the corporate environment become more selective and less visible and are *accompanied by a number of metamorphoses. Successful implementation of a management technique usually bears witness to a capacity to relativize its basic hypotheses, or even to deviate from them if necessary. It is then easier to define the indications (in the medical sense) of this type of project* and to have a better understanding, if not control, of the transformations which management techniques stimulate, whether they concern expertise or systems of relationships. But these transformations cannot easily be predicted by virtue of the initial contents of management techniques alone; they are discovered in practice, while the tensions, stakes or new structures progressively start appearing. To understand what makes management techniques both a source of "illusions"[17] and a medium for creative intervention, we have to look at their foundations: what exactly do they comprise?

3.1 "Rational myths"

We have already mentioned elsewhere[18], in relation to an idea on the role of formal models in corporate life (the example studied was that of operational research) that

---

15 Taylor F.W. (1911).
16 Plossl G.W. (1980).
17 Pavé F. (1989).
18 Hatchuel A., Molet H. (1986); Hatchuel A. (1988a).

management techniques could be qualified as *rational myths*. The concept is voluntarily paradoxical. Through it one can postulate that the objective dimension of this type of technique must necessarily be bound here to more metaphorical representations without which it would not be possible to evoke a comprehensible field of action, nor mobilize potentially interested actors. What we think we have learned from AI, in our research, confirms such a theoretical proposition and is an invitation to pursue this line of thought. Both AI and operational research constitute a source of rational myths, even if the ingredients are not the same. Furthermore, the AI movement and expert systems add an extra chapter to the history of management techniques, reinforcing an overall perspective of them and thereby emphasizing their common structural features. It is these features which, by the effect of series, we believe we can identify better now. Recent work on accounting techniques[19] or on methods of evaluating banking risks[20], which we have studied in detail, suggest that these conclusions have overall relevance.

3.2 A ternary structure

Whether they focus on human gestures (scientific management), the flow of products (computer-assisted production control), decisions (operational research), or expertise (artificial intelligence and expert systems), management techniques are modelling projects, aimed at the production of formalized knowledge. But they are not, for all that, purely scientific undertakings. Modelling is not an end in itself nor aimed exclusively at describing certain objects. It is mostly guided by potential material and relational stakes which it progressively makes credible and which each management technique had to express, represent or imagine in order to become a mobilizing project. Thus, all management techniques seem to be based on three different but interacting elements which we shall call *a technical substratum, a management philosophy and a simplified view of organizational relations*. Each of these management techniques is then a singular conglomerate, composed of each of the three elements.

3.2.1 A technical substratum

Although the technical substratum may be limited in Taylorism, it does effectively exist. The measurement and planning of actions, the elaboration of timing expertise and quasi-universal tables of elementary movements were, for many and

---

19 Ponssard J.P., Tanguy (1993).
20 Sardas J.C. (1986).

for a long time, if not the essence then at least the signature of scientific management. This arsenal of techniques was to make the fortune of some companies who, like Bedaux, were its main propagators, and until the 1940s it seemed to encapsulate the basic elements of production management expertise. With operational research, computer-aided production control or artificial intelligence, the technical substratum grew wider and more complex. Computers were mobilized very early on and became an integral part of the techniques. The era of algorithms of all kinds had begun; calculations of optimization, calculus, and automation of deductive or inductive logic were, respectively, to shape the technical features of these three movements.

Yet none of these substrata taken alone had a "management" nature as such; they all belonged to the arsenal of mathematics, logic or physics. Some were indeed developed with an exclusively managerial perspective (models of stock control are an obvious example), but that is because this type of model closely integrates the second structural element of the conglomerate, which we have called a "management philosophy", an element that would be very difficult to detect in AI languages.

3.2.2 A management philosophy

A management philosophy consists of the system of concepts that refers to the objects and objectives at which rationalization is aimed. It means that a managerial relationship can be established vis-à-vis these objects, i.e. conditioning, with variable intensity, aimed at perfection, control and selection. The archetypal objectives inscribed in management techniques also express such a relationship, with archetypal taken to mean the first and general aim which founds a management technique and gives it its singularity even if it is not necessarily pursued in the field.

Thus, scientific management is inseparable from an initial logic of increasing the productivity of human work, even if the concept can only have operative meaning in a well-defined context. *Human work* is in this context the basic source of wealth, and is considered first and foremost as a sequence of movements that can be conceived more efficiently.

Work does not play the same role in operational research, where it even disappears as an object serving a sphere in which nothing but *decision making* exists. Each decision can be optimized by finding its alternatives and selecting the most efficient ones.

Computer-aided production control, albeit a form of computerization, abandons this type of optimizing aim which is too extreme and focussed too directly on isolated decisions. It attempts to describe the industrial system as a dynamic sequence of decisions essentially oriented towards the *flow of products*. What is required from this set of decisions is a general objective for coherence. It consists less of making the best choices than of finding a suitable and understandable mode of operation.

Finally, with expert systems, choices or decisions are no longer perceived as the focal points of the technique. They are merely manifestations of existing knowledge. It is knowledge that has to be automated, just as the automation of certain types of work was formerly recommended.

Differences in these management philosophies are due to the historical contexts in which each of them was born, and to the fact that they involve actors with different positions, expertise and roles. That brings us to the third element of the triad that constitutes a management technique.

3.2.3 Simplified view of organizational relations

Management techniques would be mute and powerless if they were not defined by means of characters exemplifying the roles of a small number of actors that are described briefly, even in a caricatural way.

Managers and workers are the hallmarks of Taylorism, operational research has difficulty coming up with any protagonists other than decision makers[21], and artificial intelligence introduces a world of experts and users. Yet computer-aided production management unquestionably presents the widest abstract range of characters, since it portrays the main traditional actors of industrial firms. This is, however, achieved at the expense of the technique's scope of validity, which essentially concerns manufacturing activities. By comparison, the three other techniques owe the title of universality to the limited nature of their "primitive" scene.

Finally, all management techniques include the specialist in their particular approach: the time-keeper, the analyst, the cognitician are all characters whose names change with time, and who sometimes disappear as the technique spreads, evolves, is popularized or institutionalized as a permanent figure of certain organizations. Today, the descendants of scientific management specialists are called "planners" or "methods engineers". On the other hand, there are practically no more operational researchers, or at least actors who appear as such. Because of the changes that OR has undergone, it is somewhat difficult to identify their successors; computer specialists, organizers and managers can all claim, as much as AI specialists, to be inspired by it.

AI specialists will probably evolve in a similar way, particularly since a position that is too universal rarely resists specialization for very long. As for computer-aided production management, its promoters are as much computer specialists as they are managers of industrial units.

---

21 Roy B. (1990) provides a more detailed typology that includes more passive actors in the process.

3.3 Use of a management technique: reacting to three types of incompletion

Although management techniques or major rationalization projects are based on a combination of three distinct elements, the fact remains that those who have to implement them still have a lot to discover. The history of these projects is that of a quest in which the actors concerned, be they promoters or spectators, enthusiastic or circumspect, have to find which stakes, contents, and actors can give meaning to the technique envisaged. This helps to explain why management techniques pose a particularly difficult problem of transference. Even if a technical substratum seems to behave like an invariable and moves from one place to another like a "black box"[22], the models based on it have to be reinvented every time, as does the managerial philosophy or the organizational outlook. A management technique is implemented through an intense process of contextualization; its success proves the presence of favourable "ground" capable of successfully achieving this contextualization, far more than it does the efficiency of the technique. This same idea can be expressed by noting that a management technique is by nature triply incomplete and that, in order to observe it in action, one has to look not only at the technique itself but also at the explanation for the singular form it has assumed. The triple incompleteness of management techniques remains, in spite of the wave of myths and ambiguities that it spawns, the peculiarity of a society in which the idea of rationality is continually reborn from its own ashes – and where collective action is always uncertain.

4 The AI actor: an autarchic producer of expertise

By situating expert systems alongside other management techniques, we wanted to show that we were not faced with something fundamentally new in corporate history. The innovation created by expert systems is part of an old and constantly renewed movement of rationalization. Rationalization is taken here to mean all the efforts made to improve the processes mobilizing men and machines, with the understanding that the criterion of rationality is defined each time in a contextualized manner, with contextualized knowledge. However, even if the structure of these projects remains the same, each of its elements displays, in the case of expert systems, a considerable change of the industrial paradigm.

---

[22] We have borrowed this idea from Latour B. (1987) who used it to show how sciences and technologies circulate in society.

4 The AI actor: an autarchic producer of expertise

The most striking feature in expert systems is the *weakening of the managerial philosophy*. The project no longer aims at reconstructing or reinventing work, or even at doing better than human experts. The limit is simply placed on the partial or total automation of an intellectual or cognitive process even if, in reality, such automation may in itself be a medium for this type of reconstruction. There is also a participative aspect to the approach; no matter who and where the experts are, they will be part of the project provided that they have the relevant and automatable know-how. In this sense the contrast with operational research or scientific management is marked; with the latter the analyst was expected to find the best modus operandi possible. In fact these differences represent a profound modification of the organizational hypotheses inherent in any rationalization; here, it is the conception of the actor in a firm that is renewed.

The actors in AI are no longer the good and docile workers of scientific management, nor the cold perfectionist deciders favoured by operational research. They cannot be reduced to a black box repeatedly processing information related to their position in the network, or seen as players of go, slowly weaving their winning strategies. They are experts who dispose of a heritage of knowledge built up during the course of their work, their experience, and the positions they have occupied. This capital (the word can be taken in the economic sense here) allows them to act efficiently in a large number of situations and, even if they do not make the best choices in their particular field of intervention, very few people could make these choices for them. The automation of a part of this capital of knowledge cannot be undertaken without their active co-operation. Better still, they are the only ones capable of evaluating the quality of the automation achieved, and therefore the validity of the system's results.

Recurrent discourse on the difficulties of collecting knowledge indicates the unavoidable nature of this dependence, rather than the existence of reservations or outright and persistent resistance by the experts to this collection in the field. We have, moreover, never encountered any such cases, or only in a very mild form. They could be interpreted as the difficulty experienced in defining relevent and automatable expertise (as was the case in Naval or Cornélius), rather than as a fear of losing expertise. Moreover, such reservations may be more apparent in potential users than in the experts themselves. The latter may consider the prolonged process of exchange devoted to them by project leaders as a form of recognition which they sometimes expect.

Expert systems include a principle of co-operation which exceeds that of involvement or participation, for what is recognized in the experts is not only their aptitude to enter into a process of dialogue to express their needs, nor their autonomy in acting in their specific fields, nor a professional capital which anyone could acquire with the relevant training; it is their capacity to take advantage of the different problems that they have had to solve in order to produce their own body of knowledge.

It does not matter if this body of knowledge is heterogeneous, incomplete or inconsistent, as long as it is relevant. One could add that the AI actor is endowed with learning skills, but the expression is only appropriate where the word learning refers to an ability to acquire existing knowledge *and* to develop or produce new knowledge. The actor is thus not only endowed with autonomy but is also, more generally, capable of what can be called an *autarchic production of expertise*.

The expert system project therefore appears, in its most usual form, as a means of restoring a more collective or more controlled life to some of this expertise by computerizing it. We have already shown that this idea does not explain all the problems of modelling and developing the projects, but the fact remains that it gives specificity and a friendly tone to the rationalization process; *expert-friendly*, one could say!

5 The unexpected repercussions of rationalization: the progressive discovery of collective action

Equipped with an archetypal philosophy and a particular vision, each of the management techniques has spawned waves of applications which all provided an opportunity to experience the limits of the formulae or principles used. While the Taylorians provoked intense debate at the beginning of the century, reaction to their methods generated an increasingly rich and complex perception of work. The human relations school fed on Taylorism while trying to show its limits. It was also in the field that the initial foundations of scientific management were reviewed and revised while, during the inter-war period, Taylorians and social scientists tended towards a more relational conception of productivity.

Operational research was not to escape the same wave of reconstruction. There was a great deal of objection to the rational and omniscient decider, on the grounds of this actor's obvious limits, the multiplicity of potential centres of decision making throughout an organization, and the complexity of power struggles or negotiations between them. In short, a vision which was too naively rationalistic and focussed on a recognized leader, was rapidly opposed by a more "political" one[23].

But in each case it was the rationalization project that was thought provoking, because it over-simplified a complex reality, but also because this complexity could be organized more efficiently. The social sciences seem to have found in rationalization projects both the opportunity and the terrain for a critique directly

---

23 Abundant literature is devoted to this theme. See Simon H. (1947); Thompson J.D. (1967); Crozier M., Friedberg E. (1980); Pfeffer (1981); Mintzberg (1983).

relevant to their own development. In fact the ups and downs of operational research proved to be invaluable for an understanding of decision making systems and a renewal of sociological common sense. Until then the latter had been too inclined to neglect the uncertainties and complexity of each case and the impact of common reasoning. Operational research helped to highlight all these important elements[24]. It is similarly possible to show that most forms of rationalization had the same type of repercussion, but describing each of them here would exceed the limits of this book.

The above analysis of management techniques has several advantages. The first is of a genealogical nature and helps to support the idea that expert systems are a form of rationalization resembling all others. The second advantage lies in the fact that the analysis helps to show how experts systems have the effect of revealing hidden facts, likely to lead to a more thorough understanding of organizational life. *This is what we shall see now by examining the essential stakes in this new type of rationalization, whether they are the result of its implementation or of the unexpected repercussions that it may have.*

---

24 Hatchuel A., Moisdon J.C. (1984).

Chapter 5
Hidden crises of industrial expertise
Practical and cultural stakes in expert systems

What are the stakes involved in expert system projects? There are, of course, several ways of answering the question. One of the most classical consists of evaluating the diffusion of expert systems in the business world. Surveys published on the subject[1] generally point to a broadening of the fields of application, but provide very little information on the expertise involved or on the conditions in which this type of project is conducted. We shall therefore limit ourselves here to the cases that we studied, specifically to define the stakes associated with this form of rationalization. These stakes may be directly related to the type of industrial economy with which we are faced, or may include more doctrinal aspects because they concern a specific view of corporate functioning.

We shall first see how the projects studied reveal the crises threatening industrial expertise. Such crises mainly affect actors born of Taylorism (methods engineers, planners, maintenance technicians and so forth) as they are specifically responsible for the dynamics of knowledge in firms. This, in turn, helps to explain why some crises, the consequences of a variety economy, can be reduced or better controlled by expert systems.

Beyond these specific contexts, such analysis shifts our perception of organizational phenomena. *If we want to explore expertise, its character, its crises and its impact on intra-firm relations, we can no longer limit ourselves to a view of organizations that is too mechanistic, institutional or political.* However, if we are to pay closer attention to the processes in which expertise is formed and exchanged in practical situations, we also have to understand how bureaucratic, institutional or political processes can inhibit, distort or favour the dynamics and distribution of knowledge compatible with current economic requirements. In the light of this interaction, *should new relations between work and the production of expertise not be conceived and established?*

---

1 On the diffusion of expert systems in industry the reader is referred to O'Farell P.M., Pingry J. (1988).

1 Industrial stakes in expert systems: the hidden crises of industrial expertise

From the outset, expert systems were credited with considerable economic or industrial interest by most commentators, who based their judgement on two types of argument. The first emphasized the originality of the approach: rapidity, modularity and flexibility in the development of software were advantages provided by the new form of computerization. We have seen, however, that these properties were not achieved equally with the modelling of each type of expertise. Yet, independently of knowledge automation, such properties may well be very useful for conventional programming.

The second type of argument brings us back to corporate functioning. Some authors have seen a significant economic challenge for expert systems in the idea that modern firms have a large appetite for sophisticated expertise, but that the transmission of this expertise is still extremely costly or even impossible by conventional means. Consequently, a particular use of expert systems is emphasized: that which consists of dividing up knowledge, since the software is meant to facilitate its transfer from experts to non-experts. We have already witnessed the practical difficulties of implementing such a process, but this point does nevertheless deserve more clarification.

1.1 Conditions for transferring expertise: relevance and common knowledge

The transfer of know-how can, *a priori*, be achieved in two distinct ways: either during training, or in real work situations.

The first case is closely related to computer aided learning, the system being used outside of a working context. Although we cannot discuss all the different CAL methods here, it is nevertheless noteworthy that several expert system projects have resulted in the development of training tools, despite a more operational initial intention which encountered too many difficulties[2].

Transferring utilizable knowledge during work situations is a different and far more ambitious objective. The challenge consists of providing actors, in the framework of their functions, with a programme capable of helping them to use knowledge that was not originally their own. In the projects that we studied, this

---

2 Research undertaken by the Paris public transport system (RATP) on several expert systems showed this orientation towards a training objective; see Blanc M. et al. (1989).

type of use was a focal point in one case only – that of Cornélius – without, however, being altogether absent from the other systems. For example, the use of GESPI to test certain hypotheses on train schedules, resembled a transfer of knowledge. In the realization of such systems, modelling must be carried out from the *point of view of the user with the least knowledge* and must therefore be relevant to that user. It is, however, important to avoid the paradoxical situation in which expertise has to be transferred but also has to remain comprehensible and acceptable to a user who has no knowledge of the subject. In fact, this type of transfer implies a *minimum of shared knowledge* and the greater the amount of shared knowledge, the greater the transfer of expertise will be. For the process to have any meaning, the expertise that is transferred must have a certain degree of complexity; it must, at least, be more complicated than something that could easily be transmitted by means of a manual or a video cassette.

These remarks suggest that the use of expert systems as an instrument intended for the transfer of know-how between experts and non-experts is probably the most difficult to achieve, and that it can only be successful in a group of actors with a wealth of common knowledge.

Once these difficulties have been ironed out, we realize that the transfer of expertise, when taking place in a real work context, is also a form of training. It would be difficult to imagine someone using, for example, a maintenance support programme and not progressively "learning" from its unexpected answers, advice, or even just its way of asking questions. Such software should therefore progressively reduce the distance between the producer and the initial receiver of the knowledge. However, in conventional industrial contexts this type of use might be limited – because of the strict conditions of viability – to situations in which both expertise and its transfer are important; provided this stake remains relevant for a sufficient length of time.

Whilst this first discussion leads us to qualify the most obvious and prevalent interpretation of the stakes associated with expert systems, it also gives us an indication of method. There is less to learn by investigating the potential use of existing expertise than by looking at the characteristics of its formation and the factors determining its distribution. We shall see that with this perspective it is possible to usefully distinguish two types of knowledge evolution in an industrial context: on the one hand, the increasing complexity of "living expertise" and, on the other hand, the generation of "singular expertise".

1.2 The increasing complexity of industrial expertise: proliferation and heterogeneity

In the two cases of TOTEM and GESPI, the industrial or socio-economic background which gave meaning to the project was in some ways similar and was characterized by accelerated and unmanageable complication of activities.

The range of products processed by the workshops in which TOTEM was set up had grown considerably during the preceding few years owing to the huge increase in the use of precious metals in high-tech electronics or instrumentation. In response to these increasingly demanding and fragmented markets, the workshops were forced to multiply the number of alloys with which they worked, as well as the dimensions and physical characteristics of their products. The same was true, although in a very different context, for developments in the station where GESPI was implemented. The fast growth of traffic, accentuated differentiation of commuter and main line trains, increase in the number of exceptional schedules, and multiple commercial demands, all progressively weighed on the station planners' work.

This type of evolution is far from being peculiar to the contexts described above. *On the contrary, it represents a strong tendency in industrial mutation which dates from the early sixties. The change from an economy of reconstruction to one of renewal, the pressure to export to countries with an equivalent level of development (EEC in particular), the creation of a middle-class with differentiated tastes and life styles, are the main determinants of a variety economy, i.e. one in which goods are defined according to all viable market niches.* Diametrically opposed to a mass production philosophy, this is an economy in which each consumer or buyer explores the complex trade-offs between price, quality, delivery date or any other criteria of value, and in which individuals use their choices to express their own style of usage, life or social identity. Examples of the proliferation of variety in industrial supply abound, particularly in the mass production of consumer goods such as cars, household appliances, clothing and food (the pace at which products are renewed today is beyond all comparison with that of the sixties). Yet the repercussions of variety rapidly move upstream in the production process, impacting on the equipment purchased by industries which in turn demand flexibility and variety from their suppliers.

1.2.1 Abundance and instability of expertise

Faced with such developments, design or sales departments multiply product modifications, variety and adjustments. For all the actors involved in industrialization and production, these modifications are not only interesting ideas, *they also imply a continual and more or less significant and identifiable adjustment of pro-*

1 Industrial stakes in expert systems: the hidden crises of industrial expertise

duction protocols, materials, or logistics. Mostly, the actors resort to an accumulation of modifications and ad hoc solutions. No industrial facilities today can be sure of the exact product range they will manufacture in two or three years time and generally they have to deal with continual change. The result is a constant accumulation of practices which allow the main actors to adapt to the flow of disruptions and innovations and to maintain the principal production structures. Yet this dynamism generates its own problems which either remain partly unresolved or are dealt with by deploying all available resources to satisfy the customers' requirements. *A situation is created in which the actors' expertise is tested to the limit.*

For operators, the dynamics of the variety economy implies memorizing several different protocols, some of which are only used occasionally and unexpectedly, since the specialization of production lines is not always possible. One can easily imagine the type of accumulation of expertise that exists in such contexts of learning where each situation, with its solutions, has to be memorized. On the operators' level this memorization is possible because the sanction of expertise is often immediate; it is, however, unlikely that a coherent picture of what happens can be constituted in this way. It is in such problems that the origins of all recent participatory devices and attempts to obtain a high level of commitment from employees can be found. All these devices contribute towards the detection and control of the proliferation of practices, and thereby help to reconstruct coherent action plans.

On the other hand, for all the actors who plan and organize the production process (design, methods, planning and maintenance departments) the sanction lies elsewhere and the learning feedback loop is far longer and more complex. Questions such as "Did this adjustment or variation function well?" and "Is it as efficient as planned?" can naturally be answered more easily when the process has an element of repetitivity and when the impact of specific intervention can be evaluated. But with a continuous flow of modifications, the risks of opacity and inaccurate interpretations multiply. Furthermore, notions as natural as productivity, yield, or workshop capacity become complex and variable; it is then necessary to talk in abstract terms of thousands of work hours, or of equivalent products, when in fact these will obviously vary from one month to another.

1.2.2 Heterogeneity of expertise

Variety also has another, often under-estimated, consequence: the increased heterogeneity of each actor's expertise. For *variety also means the entry of a dynamic environment into the corporate productive system*. When an activity is repetitive, knowledge of the production process seems to be an autonomous and homogeneous body of know-how. Expertise does not necessarily display, once it has been stabilized, the economic or commercial hypotheses that were involved in its elaboration. Yet, when each variation of the product corresponds to different economic and commercial contexts, this knowledge becomes an integral part of know-how. For example, a certain aspect must receive particular attention for customer A, and a different one for customer B; the quotation for X was calculated with extreme care; order Y only uses components from C, and so forth. With variety it is the entire socio-economic history of the products, the way they are consumed or used, their aesthetic or sensory impact, and not only their technical history, which tend to be needed by the different industrial actors, manufacturers or designers. Moreover, these histories are themselves heterogeneous. The management of a large station, a borderline case of this type of evolution, is – as we have seen – a veritable industrial, social and commercial imbroglio.

The heterogeneity described above is manifest in the knowledge bases of the different systems studied, and particularly in three of them. Only Cornélius partly escapes from it owing to the relative specialization of the machining unit. This is not, however, a general property of all maintenance activities; for example, problems in maintaining equipment situated in very varied socio-economic contexts (domestic devices or those exposed to public use, for example) are particularly varied.

Heterogeneity, variety, instability, or sedimentation of practices can affect both high-tech and traditional industries. They give rise to permanent tension and to endemic or latent crises. Their signs are not always acute or visible; production may progress chaotically, but it progresses; some orders create serious problems, but they remain rare. On certain work stations which suffer from a high turnover, passive resistance to innovation tends to set in, for any additional change threatens a balance that is seen as both tiring and fragile. Situations of this nature are particularly awkward for managers who sense increasing opacity in the functioning of their firm but who cannot see any specific target on which they could focus their attention to take action. Although the caricatural portrait does not entirely correspond to any of the situations that we analyzed, each of them displayed some aspects of it. In such contexts the stakes in the expert system emerge as the work of dissecting and representing knowledge sheds direct light on the detailed structure of expertise. Modelling can help to partly reduce the crisis in two different ways.

– The structuring and conceptualization required in automation can make it possible to find general manipulable patterns where these were not apparent, or where they were used intuitively by experts. Such an effort is only relevant if the knowledge ordered in this way is lasting and if the accumulation of practices represents business developments. *This is what we shall call "living knowledge" and in this case the project resembles the elaboration of a more coherent engineering of the activities concerned, which could partly serve as a new common culture.*

- The modelling process can also highlight unsuspected constraints or costly recipes which have survived by sedimentation without it being noticed that they were no longer valid in a new context. It is thus possible to reveal the constraints weighing on certain actors, who quietly take responsibility for choices and compromises which unite all the contradictions to which the activity is subjected.

In view of the consequences of variety, and particularly the heterogeneous and proliferous increase of practices to which it can lead, there are multiple major stakes involved in the implementation of expert systems. But to achieve its objectives, the project has to include a re-evaluation of the situations experienced by the experts, as well as an effort at recomposing a new area of expertise and new relationships. It would be absurd to claim that such recomposition necessarily involved an expert system project. Yet, any other method would imply the same type of discovery and action with respect to expertise and practices, and would also have to be based on mobilizing principles and active promoters in the firm. The development of an expert system is only one of the possible mediums for this mobilization, although it is one which managed to unite, in the cases studied, the necessary enthusiasm and resources.

Industrial life is not, however, composed only of prolific developments; it also has its ruptures and its imbalances.

1.3 The effects of technological and economic imbalances: singular expertise

Besides the processes of complication by increased accumulation and variety, firms also experience ruptures and imbalances: markets disappear as quickly as they appear, production processes do not always meet expectations, and so forth. These crises can destroy an activity, but they can also be overcome. In such contexts, certain actors develop what can be called *"singular"* expertise – knowledge which grows from the necessity to deal with sudden problems and to stick to the situation. Here action is the only form of learning. This type of expertise may develop when a technology is not mastered or can only be used by a small number of persons; it may also be found in contexts where difficult negotiations take place, where only very few people involved have any experience. The need for such expertise can last as long as no balancing process is initiated: Naval and Cornélius were born of this type of constant imbalance. Cornélius was an attempt to solve the problem posed by an unreliable facility with which very few people were familiar, whilst Naval had to find acceptable compromises in the face of contradictions created by shocks on the oilrig market.

In so far as it may help to organize the preservation or the transfer of rare expertise, the expert system is seen here as a means to reduce an existing imbalance. In fact, besides the problems already mentioned concerning the transfer of expertise, it is particularly relevant to wonder whether, faced with such singular expertise, automation really is the best strategy for dealing with an imbalance. Due to the difficulties and considerable amount of time involved, does the project not risk arriving too late, to find that the imbalance has either disappeared or worsened? Are there no surer alternatives? It is noteworthy that, amongst the projects studied, those which did not survive (Cornélius and Naval) became irrelevant because the same problems that had engendered them were solved or replaced by a totally different context.

Whilst the relevance of an expert system project should not be systematically excluded in this type of context, the project should be assigned specific objectives. In Cornélius and Naval, the process of revealing and systematically modelling expertise had its positive repercussions: Cornélius seems to have facilitated the methodical organization of improvements, making the facility more reliable, whilst Naval shed light on existing modes of managing crises on the oilrig market. In this type of situation, the project's main value lies in what can be learned from it to help reduce imbalances or select an appropriate attitude vis-à-vis these imbalances. The fact remains that such repercussions are cast aside or are inoperative if the imbalances persist or continue to shift.

Thus, it is by better understanding the industrial dynamics most peculiar to our times that we can define the main challenges which can be met by expert systems. *We are inclined to think that the main impact of such projects is their contribution to the identification and reduction of crises due to the variety economy.* To sum up this conclusion, one could say that whilst automating an expert's knowledge is a good thing, helping an expert to reformulate and exchange his or her expertise more effectively is probably more valuable.

Finally, these experiences help us to discover the problems at the heart of industrial dynamics today, mainly because the expertise in question here belongs to actors born with Taylorism, or at least in the form this movement took at the start of the twentieth century.

1.4 Mutation of planning and design expertise and management tools

By sketching a portrait of the experts asked to reveal their knowledge in the four projects studied, we noted that these actors owed their place in the corporate structure to Taylorism. One tends to forget that Taylorism, which aimed at rationalizing labour, had the specific effect of spawning several actors whose role was precisely that of defining, standardizing and organizing the productive system in which labour could best be utilized. Theirs was therefore planning and design expertise, or what we call "conception expertise" – be this product design, production system design, or planning of the daily organization of flows and teams. The know-how needed by these actors was elaborated gradually, but it was all the more valid and legitimate when variety and imbalances were limited. *Today the planner and designer's expertise, faced with the accelerated pace of innovation and composed as much of technological innovations as of the anticipation of changing market demands, can no longer be considered as absolute. It is, in turn, entering into the critical cycle of industrial reactivity.* Two requirements result: the reshaping of industrial cultures and a new role for management systems.

1.4.1 Reshaping industrial cultures

Today, more than ever, the designer or the planner's expertise resembles a gamble. He or she has to deal with technological or commercial risks which, even when calculated, are often irreducible. Furthermore, designers and planners cannot predict all the problems and adjustments that production or operations actors will have to deal with in the field. Hence, a new industrial culture common to all these actors has become necessary for two reasons. First, to enable planners to gain insight into problems in the field, or at least to imagine flexible solutions capable of evolving in action. Secondly, to allow those on the ground to perceive the gambles taken by the planners in order to react more quickly and more effectively when things deviate too far from planned scenarios.

The designer's course of action is, however, no longer outside of production time. The pace at which new products have to be marketed is so crucial to commercial success that the designer's ability to react quickly conditions the overall response of the firm. Moreover, designers and planners are no longer an inexpensive means of regulating the production process. *They are now crucial elements in a global production system where survival depends on the efficiency with which the complete life-cycle of a product is managed.* It is therefore not surprising that present rationalization strategies also focus on these actors and their main instruments – their expertise. Nor is it surprising that these actors were naturally designated as experts in the industrial system.

Thus, the key elements of a mutation and redistribution of "conception" expertise emerge. These are:

– *the use of automation* when possible, which in itself necessitates a reformulation of conception expertise[3];

---

3 The automation of planning expertise has several forms: the development of CAD or CAM tools naturally constitutes a significant axis of this process for research depart-

- *the taking into account of the inevitable heterogeneity of this expertise.* Expertise applied to practical situations is not an academic discipline; it is a product of the integration of multiple levels of knowledge. This heterogeneity can be managed by the actor in his or her own process of producing knowledge, but with the permanent risk of haphazard accumulation. It can also be regulated by a process of reconceptualization and restructuring which makes it more intelligible even if the result is not automation. Considering the multiple problems posed by interchange between designers[4], heterogeneity cannot be reduced to the mere co-operation of several actors with specialized fields of expertise, unless they share a strong industrial culture;
- *the reconstitution of knowledge shared between planners or designers and operators, leading to a partial redistribution of expertise.*

Such transition does not take place easily and while different forms of this evolution can be witnessed in many cases, it remains dependent on the nature of knowledge shared by the relevant actors. *Whatever the pace at which firms evolve, it is probably around this very question that the most important stakes are to be found.*

1.4.2 A new role for management tools

The mutation of conception expertise has an impact on the main management techniques.

To manage any activity, one needs a representation of the activity, which must comply with the following conditions. It must:

- be an acceptable description of the phenomena;
- allow for relevant strategic interpretation and action;
- help to predict the most significant effects of this action.

A manager's expertise therefore requires the three types of know-how analyzed in Chapter Two; he or she must have at least some elements of "doing know-how", "understanding know-how" and "combining know-how" concerning the activity to be conducted or controlled. In fact managers draw upon all kinds of conception expertise, and it is this expertise which supplies them with the main variables of the activity, or with possible strategies. *When conception expertise is stable and valid, the representations which it provides can be used as a prediction and therefore as a standard.* Productivity, estimated costs, production, sales and so on, can

---

ments, but on a technical and philosophical level it is different from expert systems. The reader is referred to Poitou J.P. (1984) for a history of the evolution of design departments in the automobile industry.
4 David A., Giordano A. (1990).

1 Industrial stakes in expert systems: the hidden crises of industrial expertise

be assigned target values because the means to attain them are known. Deviation from the norm is accepted but considered as residual, the fruit of imponderables.

If, however, we supposed that such predictions were not possible and that we could express targets, but without being able to state exactly how they could be attained or what obstacles might arise, *management techniques would not produce norms, but rather instruments for coordinating and monitoring, for setting a course as a guideline.*

The adoption of either philosophy results essentially from the diagnosis that is made of the state of conception expertise, its contents and its validity. Such diagnosis is closely linked to the epistemological evaluation that actors make of different available expertise. For example, what is a machine's nominal performance, stated by its manufacturer, worth to a workshop supervisor? Would he accept these values *a priori* in a definition of his mission, considering all things as being equal between his workshop and the supplier's test bench? Or might he ask that these values be taken simply as a benchmark, of which the feasibility will be discovered step by step? We see that these values will or will not be included in an expert system generating this workshop's routings, depending on which alternative is chosen.

Hence, recent research on management control or industrial accounting (like activity based costing), rather than considering[5] novel management concepts, contributes to a new vision of the usage and interpretation of all business models, where conception expertise no longer provides standards but defines targets.

The organizational consequences of this vision are no less important, for with it relations between staff controllers and line managers have to be rethought. *The essential mission of controllers must then be to provide line managers with the means to control and co-ordinate, which the latter will be in a better position to interpret and analyse considering the objectives assigned to them*[6]. The difficulties of this type of evolution are multiple. In the long run, will there be an acceptable position or "actor figure" for the staff controller? Probably, as long as the establishment of management norms on certain questions (other than definable objectives on products) appear legitimate; this is still the case with quantitative employee levels in many industrial situations.

All these elements help us to understand why the recomposition of expertise is probably the essential stake in current managerial and industrial dynamics. But it is at least equally important to emphasize the fact that this recomposition can be a new form of rationalization (e.g. a new form of scientific management) without retaining the bureaucratic features of classical Taylorism:

---

5 These long-standing problems are the subject of a number of current studies. Many of them refer to recent American attempts to renew management control. On this point see C. Berliner, J. Brimson (1989); Fixari D., Hatchuel A. (1990); P. Lorino (1990).

6 This analysis was taken further in an operative and suggestive manner in Ponssard J.P., Tanguy H. (1993).

- In effect, formal modelling is necessary for the recomposition of planning expertise, and it requires a very active Taylorian attitude on all levels. Firms have to carefully review all their expertise and practices and to make use of measurements and formalism wherever they seem relevant.
- This attitude no longer necessarily leads to a clear separation between conception and execution, considering the limits of validity imposed on conception expertise by a variety economy. It makes it possible to better identify the gaps and risks inherent in any planning or design work. One can then redefine the type of co-operation between conception and execution by which the potential consequences of these difficulties can be limited or controlled in the field. We are thus faced with the progressive establishment of what can be called "*participative Taylorism*"[7]. Does this double nature of the current process not lead to a "straddling" strategy or to a painful contradiction?[8] This may well be so, but there is nothing fatal about such deviation. Avoiding it depends precisely on the mode of recomposition of expertise between conception (or planning) and execution, and the fact that the "executors" do not limit themselves to mobilizing their own expertise but are also actors in the rationalization process.

Conception work perhaps remains the main orchestrator in the life of the productive system, yet it now has to deal with new co-operative links on the value chain, links which will trace the obscure paths of economic survival.

2 Firms as producers and legitimizers of expertise

This glimpse of the rationalization of industrial expertise leads us to more essential and less familiar questions: *how is expertise formed in organizations? How is it recognized as such, and how is it validated or legitimized?*

In the preceding chapters we noticed more than once how systems of relations (roles, hierarchies, communications, territories) and systems of expertise can either oppose or mutually support one another. Thus, the presence of a planning department helps to institute the existence of planning expertise, even if the AI specialist only manages to record a few bits of automatable knowledge. Or, the establishment of an additional level of abstraction in a hierarchy of knowledge can legitimize an organization chart with a similar structure. These phenomena do not necessarily correspond to established professional logics. Even if the latter enjoy a high level of importance, and are one of the social modalities legitimizing knowledge, they do not necessarily merge with the dynamics of expertise in col-

---

7   Hatchuel A. (1988b).
8   As D. Linhart suggests (1991).

lective action. Thus, the status of engineers in firms does not tell us much about the knowledge mobilized by an engineer in a particular industrial situation.

We are then led to interpret organizational phenomena not only as a social mechanism which is partly produced by instituted or recognized expertise, but also as one which in turn structures or produces expertise. The dynamics of the expert system project usually make such a reinterpretation necessary; *by wanting to explore an expert's know-how, we also reveal the collective conditions of the production of this know-how.* From the point of view of corporate evolution, two questions result from such an observtion: which are the systems of relationships that favour the production of relevant expertise and, conversely, what do systems of knowledge teach us about systems of relationships?

It is of course not possible here to try to answer such questions fully, but we can use the experiences analysed to put forward two main lines of thought.

2.1 Training processes: new relations between work and learning

The first line of thought, which we shall mention only briefly here, refers to the training processes which play an increasingly significant role in corporate advancement policies.

The study of expertise in action shows that it is not formed by the mere acquisition of knowledge, but by the reformulation of elements of knowledge in a practical context. *There is a reconstruction of a new form of expertise within a relational system, which gives this system its relevance.* The acquisition of new skills would therefore be accomplished only at the end of this reconstruction. This explains why training can only be a single phase in a wider process of transformation aimed at recreating a new productive system. On several occasions during the projects studied, knowledge was transferred from one actor to another, from the expert to the planners and vice-versa, *but this appeared less as specific training than as the elements of co-operation essential to the projects underway.*

The need to bring training closer to the process of action and innovation has already been dealt with on several occasions and has resulted in experiments in "action-training". But there is certainly room for the development of training processes based on more specific thought on different types of expertise and the conditions in which expertise is elaborated in action. An analysis of several experiences in retraining unskilled workers has shown that, in order to be successful, such operations have to be part of an effective process of industrial improvement. Only then is it possible to identify the expertise that it would be useful to transmit or to produce. This model is not very different from innovation schemes which discover their own phases and the actors they need as they progress[9]. On this basis it has been possible to set up experiments in training which alternated the provision of knowledge, on the one hand, with the involvement in real projects started by the participants in their work, on the other hand[10].

It is thus clear that the complexity and pace of innovation in industrial life are not compatible with training that is too far removed from the real-world situations. At the same time, they do not lend themselves to learning on-the-job either, beause there is too much information to integrate and too much variety to assimilate alone. Firms will have to mix proximity to real situations, which is a guarantee of relevance, with the acquisition of knowledge and conceptual thought, which ensures overall intelligibility.

Whilst corporate functioning relies more and more on the maintenance of a permanent process of production and legitimization of new knowledge, it is necessary to ensure that training (acquisition of relevant and coherent knowledge), work (implementation of knowledge within a system of relations) and research (production of new knowledge) are structural elements of any activity.

2.2 Weakening of hierarchical relations and dynamics of expertise

The importance today of understanding the ties between systems of expertise and systems of relationships, lies in the fact that this interaction becomes essential when hierarchical relationships weaken in large sections of organized action, giving way to more horizontal relationships. *One could even add that such horizontal relationships are possible only if there is permanent rationalization of expertise, which maintains concepts and knowledge common to all the actors concerned.* Otherwise how could actors who rely on different expertise and who work in complementary, although not necessarily disjoint, areas of intervention, co-operate? How could they renegotiate their frontiers as the situation evolves? Such horizontal relationships are formed, for example, during the development phase of expert systems. In the cases we studied, the project leaders – in spite of their belonging to different professional levels – had to occupy a position of intense observation and exchange for many months, with persons who were hierarchically below them. The observed result was that, at least during certain periods, institutional functions became indistinct and relationships oriented towards the collation of knowledge were established.

---

9 Fixari D. et al. (1991).
10 Dubar C. (1991).

2 Firms as producers and legitimizers of expertise

The stakes related to increasingly horizontal corporate functioning lie here, for even a principle as prevalent today as "client-supplier" relations requires considerable effort to collate knowledge when the context is no longer one in which standardized and stable goods are exchanged. Recent research[11] on contracts in an unstable environment shows the important role of the contract as a guide for action. It also emphasizes the need to regularly revise a sharing of knowledge and a common vision to avoid the rupture or incoherence of behavioural patterns.

This brings us to one of the main points of our analysis: *the collation of knowledge is an activity which produces a wealth of relational models*. It reveals potential similarities between different systems of knowledge or forms of expertise; it highlights common or distinct objects and concepts and shows areas of incoherence or unsuspected discrepancies[12]. One can thus hypothesize that a better understanding of the conditions in which expertise is formed will constitute a source of new relational paradigms, increasingly needed for the intelligibility and even the production of collective action.

After all, our concepts in this respect are not very rich and have often shown signs of being inadequate. For example, the classical principles of hierarchy, work division, territorial partitioning, or technical co-operation, in no way account for the complexity of the life of planning teams[13]. It is also significant that in one of the most elaborate and prevalent organizational theories, most firms which innovate or mobilize a large number of experts are defined as "adhocracies" whose main co-ordinating principle is "mutual adjustment"[14]. This is, of course, an elegant way of saying that their functioning is not understood or does not correspond to any common principle.

But without a reconstitution of the dynamics of the expertise concerned, it would be impossible to understand much about the projects that we studied. By identifying these dynamics, we at least have a guideline to better comprehend the relationships and notably the functioning of organizations around a project, since the peculiarity of such organizations is precisely that of experiencing a continual adjustment of expertise.

It is also by investigating systems of knowledge that we can approach, and possibly get beyond, some of the usual antinomies of organizations, particularly that of *locality and globality*. The property of an element of knowledge is obviously that of being able to belong to several actors simultaneously, *while also being part of different lines of reasoning or frameworks of knowledge*. An example is the planned date for the end of a project, which is known to all the participants but used differently by each one. This may be a trivial example, but the mechanism would be the same if, instead of just a date, we considered more complex

---

11 Ponssard J.P., Tanguy H. (1993).
12 David A., Giordano A. (1990).
13 Moisdon J.C., Weil B. (1991).
14 Mintzberg H. (1983).

blocks of knowledge such as the reasons behind the launching of a project, or the main types of constraint encountered by certain actors. It is this *shared knowledge* which can regulate the divergence of different actors' viewpoints, not by realizing a total fusion of objectives or removing all tensions and conflicts, but by making each actor's intervention intelligible.

Artificial intelligence formulates a project for the automation of expertise in a caricatural manner; it can therefore, by inversing this perspective (a common mechanism in the history of rationalization) prompt us not only to evaluate its hypotheses, but also to analyse organizations by paying more attention to actors' expertise and to its distribution. This type of research could offset theories which are too structuralist or focus too extensively on actors' strategic games, since they all consider the creation of expertise as being self-evident.

It is, however, rather its ability to give a clearer and more relevant outlook on current industrial questions, than its theoretical interest, which could justify a new orientation in research. *Thus, on what principles should systems of relationships be based, and according to what correlative principles should they be understood, when the division of work disappears into the mist of participative or horizontal mechanisms?* Owing to their natural ubiquity, the infinity of their connections and metamorphoses and the variety of their structures, the study of knowledge systems and their modes of exchange can help us to understand what we might otherwise see as "disorganized" because our organizational models are too poor; for in organizations, like any other phenomenon, we can only read the patterns that we are able to conceive. Evidently, such an approach does not exclude the relevance of other points of view, it does not remove conflicts of interest, or cultural influences, it simply requires that we think about their links with modes of production, validation and distribution of expertise.

We are convinced that a new vision of corporate actors is inscribed in the rationalization of expertise, just as the seeds of the human relations movement lay in the rise of scientific management.

Chapter 6
Conclusion

Let us retrace our main lines of thought. We followed four expert system projects and, as a start, had to look at the core of the expertise that was to be modelled. This helped us to see that the basic hypotheses of expert systems were transgressed whenever the expertise in question differed from a particular model, i.e. that of elementary knowledge which can be linked by universal rules of logic. It was thus necessary to reconstruct a part of the reasoning and to reshape the pattern of action when faced with diagnostic or compromise expertise.

We then discovered the significant task of transforming expertise which had been accomplished, and identified exactly who the experts were and how their expertise was formed. It was thereby possible to distinguish several types of expert, in terms of to the balance between the recognition of their expertise and the positions they held.

It was this balance which was progressively shifted by the production and redistribution of expertise during the project, leading to the transformation of the actors concerned. This process of change, which we have called the *metamorphosis of actors*, originated in the production of new knowledge. By analyzing it, we have attempted to enhance classical organization theories which are ill-suited to the life of firms in a context of constant renewal of expertise.

Nevertheless, in spite of their originality, expert systems correspond to an older form of rationalization. By taking a more historical view it is easier to see what they have in common with scientific management or operational research; and we find the same main structural elements at the base of these management techniques: a technical substratum, a philosophy of efficiency, and an organizational model.

Each of these successive movements has created a *rational myth* which calls for change and reveals some neglected aspects of corporate functioning. Expert systems are no exception to the rule. The expertise to be modelled is *"conception" expertise* belonging to actors born of Taylorism ("thinking" departments, workshop planners, maintenance technicians) and this process sometimes uncovers hidden crises which these actors have to cope with in today's *variety economy* context. Even though the expert system project can help to alleviate these crises by provoking a recomposition of this expertise, the latter still has to be "living", in other words, it has to develop by accumulation and proliferation. On the other hand, in the face of the ruptures and imbalances of such an economy, it is rather with the production of entirely new expertise that firms are confronted.

But, like preceding rationalization movements, expert systems will progressively fade into the background. In view of the complexity of situations and the allusive nature of initial concepts, several different paths will probably be explored – and some are already, since we can witness a new generation of more open languages. Thus, the initial basic hypotheses will probably break down and, depending on the situation, emphasis will be placed on knowledge or on a more flexible and user-friendly type of computing. We will probably talk less in terms of expertise, know-how or imitation, as more abstract and more technical concepts appear which will simply state that we know how to process certain particular structures of information. Once this stage has been reached the method will have become commonplace; what was a suggestive and somewhat vague project will become, more prosaically, a technology with too many uses to be defined easily.

Nevertheless, everything that we were able to learn from this rationalization, on the nature of expertise in corporate activities, on the status of experts, and on the current crises in design and planning activities, remains relevant. It therefore seems to us useful to conclude by looking at three of these: *the problems of automating expertise, the importance of a new conception of rationalization, and the stakes of a better sharing of knowledge.*

1 Problems of automating expertise

Expert systems are probably only a stage in the history of the automation of expertise, but their implications will certainly be just as valid for later phases. First, it will still be necessary to understand in each case the particular interdependence between knowledge and way it is processed. Like with expert systems, it will not be possible to preclude more thorough reflection on the different forms of expertise, and rather than looking for a hypothetical art of questioning experts, it is likely that we shall witness an integration of other modelling traditions (with, notably, branches of operational research).

But, most importantly, there is no doubt that this automation will necessarily be a reconstruction or a transformation of the expertise that is to be automated; it will be a new form of expertise, which is neither an abstract synthesis, nor a collection of existing theses, but *a systemized set of answers to a range of problems.* Such a reconstruction will only be relevant to certain specific organizational contexts, and the choice of experts who participate in them will itself be a way of defining the new expertise. There will always be a question mark on what makes such expertise acceptable and on the durability of this validity. Yet the project will not necessarily end with the development of an automat, for all its repercussions have to be taken into account even if this means abandoning the very idea of an automat. This view is hardly surprising, if we have a clear perception of the general structure of rationalization processes.

2 A more accurate perception of rationalization

As long as the idea of efficiency has meaning, the projects undertaken to set up more effective productive systems will be perceived as a form of rationalization. Of course, all acts in economic life could fall into this category, but when an organization becomes too complex to be totally thinkable by a single man, the transformations to be made can no longer be described in the detail of each elementary act and are *shaped along broad rationalization models*. If it is conducted efficiently, this movement is inevitably a process of discovery, during which the stakes, the actors and the targets of transformation all develop together. The current complexity of economic processes and their instability only intensify the importance of this open management of the process, since the new possibilities discovered along the way count as much as the initial goals of the project. The automation of expertise is no exception; it is better to produce efficient expertise than to automate obsolete expertise or to have to manage an automat containing the recipes of an expert who is no longer around. Moreover, the production of new expertise may well be an outcome of efforts at automation. *Nevertheless, it is not possible to manage such projects without being consistent and careful to maintain an understanding, amongst all the actors concerned, of the successive goals of the approach and the metamorphoses to which it gives rise.*

3 The stakes of a better sharing of knowledge

The automation of expertise, as we know it today, is part of the present efforts of many industries to reconstruct new planning and design expertise adapted to the shocks of the variety economy. At the start of the nineteenth century, when running a business was limited by the necessity to add and multiply large numbers of figures very quickly, it was arithmetic that was automated. But expertise in a variety economy is not that simple; it is complex, due to the accumulation of objects and functions, it is also heterogeneous, the fruit of multiple technical, commercial and financial compromises and, finally, it is sometimes unstable and can be renewed unexpectedly. Automation can only concern a part of this expertise, and firms will have to use more creative resources to resist the impact of this complexity and heterogeneity on their functioning.

Faced with such turbulence, it seems that two tendencies appear: on the one hand, the fragmentation of big organizations into networks of autonomous units specializing in types of services or products and, on the other hand, a more horizontal structuring of relations within these autonomous units. This partitioning is intended to promote and recreate better structured communication, both between autonomous units and within them. But, the problem nevertheless remains the same: faced with the complexity and heterogeneity of expertise, communication is difficult without a large base of shared knowledge, and without specific efforts to maintain this common base.

It is of course natural to struggle against complexity or uncertainty, when this is possible. *But once they exist there are not many options other than the sharing of expertise allowing each actor, not only to do what he or she is asked to, but also to be ready to react to the unexpected and, better still, to understand the consequences of the unexpected for his or her partners.*

By adopting this point of view, we seem to be moving away from research trends advocating strong corporate cultures based on shared values. However, since a discussion on the question would be incongruous at the end of this book we shall confine ourselves to stating our point of view by means of two remarks and a short example.

We note, first, that the sharing of knowledge can be posed as a value of integration and promotion. We might also add that, in the heat of action, the actors who are closest to the ground quickly grasp its practical necessity, even if this is not confirmed by their superiors. Let us now look at the example[1].

During training in a factory preparing a new assembly line, the workers contest the time allocated, the pace and the organization of operations. This is a problem that has been around for at least a hundred years. Taylorians in the last century (at least the most serious of them) would not have rejected these grievances, they would have taken new measurements and redone all their calculations, before informing the workers of their conclusions, and perhaps even agreeing with them.

What attitude would be meaningful today? Job design and scheduling is more complex and has to take into account strong variations in the goods to be produced, as well as diverse incidents. In spite of this complexity, the managers have chosen to explain some of these calculations to the workers and have experimented in making them responsible for recording the measurements and data needed for these evaluations. The debate has been completely transformed. The workers have used their knowledge of the work stations to come up with some more effective configurations, and have asked the methods engineers to investigate some solutions that were beyond their computational abilities. We can well believe that

---

1 Fixari D. et al. (1991).

3 The stakes of a better sharing of knowledge 125

the launching of the new mode of production and organization has posed fewer problems than usual. We also know that tension has not, for all that, disappeared and that other conflicts will reappear. *But new knowledge has effectively been produced, for each of these actors taken separately would never have imagined the final configuration, and it could never have been obtained by negotiation alone.*

Is there a moral to this story? No, for its true sanction lies in the efficiency acquired in the field. *We have to be satisfied with the idea that after this type of experience, the expert is no longer quite the same, and nor is the system.*

Part 2
Four case histories of expert systems

Introduction

In this section we provide readers with the case history of the four expert system projects on which the analyses and theories developed in the former section were based. There are two reasons for us having chosen to do so. The first is the relative scarcity of available literature on the subject – very few detailed descriptions have been made of the different aspects of experiments in industrial expert systems. The second is to enable readers to form their own interpretation from these four accounts which in particular provide further details on the technical functioning of the expert systems. *We could finally invoke a more methodological argument: since a systematic check on our facts would be too problematical, readers can, through these case histories, see how we studied these projects, which aspects we concentrated on, which type of detail we noted, and so forth.*

To make these accounts easier to read and to compare, we have dealt with each case in exactly the same manner. The first section introduces the problem which led to an expert system project and presents the context in which the system was to operate. The second section describes the activities of the experts involved in the project. The third section is devoted to the design of the expert system itself and to the progress of the project, with an emphasis on knowledge modelling. Finally, the last section describes the itinerary and conclusion of the project. It highlights the stakes which in the end proved to be decisive and attempts to show how the initial objectives were shifted or transformed by the dynamics of the project.

Chapter 1
TOTEM
The reconstruction of production planners' expertise

1 Problem and context: metallurgy and variety

In many metal processing industries production routings, which describe the successive production steps to be carried out, schedule work on the shop floor. The increase in the variety of orders, and therefore of routings, forces firms to find ways to automate the preparation of these documents. In the following text we discuss the case history of a medium-sized firm which chose to develop an expert system as a solution to this problem. In order to understand this firm's reasons for undertaking such an ambitious project in 1987, we shall first present the main characteristics of its activities. This will lead us to introduce the notion of routing and the different ways in which production planners prepare and manage routings. We shall then put the project back into the broader context of the profound transformation which the firm wished to make to its production system and related management techniques, in order to adapt rapidly and efficiently to a fast-changing environment.

1.1 Processing precious metals: consequences of diversity

The firm studied carries out all types of metallurgical operations (rolling, wire drawing, etc.) to transform ingots or billets of precious-metal alloys into strips, plates, wire, and so forth. It differs from other firms in the sector mainly because of the high value and small quantities of raw materials processed, as well as the reduced size of its factory. (Its biggest rolling mill is no larger than that of a major aluminium processing company's research laboratory.) The firm only manufactures to order, and for clients with vastly different requirements. There is, for example, very little in common between a jeweler's occasional order for a small quantity of gold, and the big and regular orders for micro-profiled metal from large electrotechnical firms, to be used for the mass production of electrical switches; or between the production of soldering rods or rings, used in large quantities in heating systems, and that of platinum dies which are extremely complex and expensive and are produced singly for the glass industry. On the whole, orders are not only heterogeneous (forms and usage vary, over a hundred alloys exist, quantities range from a few hundred grammes to a tonne) but generally also have specific characteristics which create a new problem every time. Out of the 20 000 orders handled in a single year, more than half are unique.

In order to cope with this diversity of orders the factory is divided into departments, each specialized in the processing of alloys based on one type of precious metal: gold, silver or platinum. Within these departments workshops are organized according to the different technologies concerned. For example, the silver department has a rolling workshop with the various rolling mills and most of the annealing furnaces and shears, a wire drawing workshop with all the wire drawing machines, and a micro-profile workshop. A single order is rarely processed in different workshops. In the same workshop it may however undergo several operations which follow one another in a relatively complex fashion, with some overlapping. In spite of specialization each workshop is faced with the extensive diversity of the orders it receives.

1.2 The role of production planners and routing

Each order presents a particular problem by virtue of its dimensions, type of alloy, state of the metal and mechanical properties. In each case the form and quantity of the required raw material must be specified, as well as the sequence of operations and the machines best suited to achieving the desired result. All this information is contained in a single document, the production routing, which plays a central role in all manufacturing industries and particularly in situations where, like here, the nature of operations and their sequence differ from one order to another. In a single workshop two orders can be routed differently, and this may either involve the same machine but at different stages of the production cycle, or different machines. The routing defines the order's route in the workshop and the work to be carried out at each stage.

Traditionally, production planners are responsible for preparing and managing production routing. These specialized technicians apply a variety of skills such as metallurgical techniques or knowledge about the machines and about workshop practices and problems. They co-operate with a number of other actors in the firm: the sales department which gives them the orders, the planning department which gives them production schedules and the workshop managers who inform them

1 Problem and context: metallurgy and variety

about the stocks of intermediary products, allowing them to select a "basic metal" (the metal that is used to start with).

Production planners adopt different work methods, depending on the firm. Three types of situation exist:

- *The operation of a library* in which production routings are archived. When a new order arrives, the production planner checks whether or not the same order has already been processed before. If so he or she uses the corresponding routing; otherwise, a similar routing must be found and adapted accordingly. This method soon reaches its limits if the number of different orders increases significantly. Routing libraries become so large that they are difficult to manage and to keep up to date.
- *The completion of standard routings* which describe, for each family of products, the main steps of the manufacturing process. As soon as the production planners receive an order, they match it to a family and complete the corresponding standard routing. It is not however always possible to distinguish families of products and to match them to routings capable of accurately representing the diversity of orders. Because of this diversity the planner may have to choose between either multiplying the number of standard routings, or else increasing the amount of work involved in completing them. The latter may finally result in the preparation of a new document every time.
- *The automatic generation of routings.* Another method is that of using the production manager's expertise to prepare an entirely new routing for each product. This work can be automated, in which case expertise must be formalized and included in a computer programme. In spite of a large number of attempts to do so, such projects have until recently encountered major difficulties and have resulted in only limited applications. Expert systems, because of their ability to present production planners' expertise in the form of rules, rapidly appeared as a promising solution to this type of problem.

1.3 Emergence of the project: the goals

The firm we studied opted for the latter type of approach. In order to cope with increasingly diverse orders and elaborate products, the goal assigned to the project was that of avoiding very large routing libraries by formalizing expertise in the form of rules. These rules would then be used by the expert system to prepare routings, particularly for new products which were not part of the workshop's previous experience. The project also aimed at facilitating the up-dating of expertise, making it more homogeneous and favouring its diffusion.

Three basic reasons, all part of more general preoccupations of the firm, motivated the decision to launch the project. First, the tightening of economic constraints (notably as a result of the internationalization of markets, the dynamics of competition, increased quality standards, and the proliferation of product variety); secondly, the new importance given to optimum manufacturing techniques (faced with the increase in the price of precious metals, customers tried to limit as far as possible the quantities of metals used, which implied additional manufacturing constraints. For example, the thickness of a strip of precious metal used in electronics was reduced by a factor of 10); and lastly, the evolution of the type of customer which contributed to a profound transformation of trades (industrial customers now represent the greater part of activities in this sector, rather than the art-related trades which were their traditional customers).

In order to adapt to these changes in their environment, corporate managers had to face the urgent question of the accelerated transformation of an industry by means of relatively stable techniques, machines and management methods. Several projects were undertaken in that direction. For example, a new micro-profile workshop was designed with a clean room in which the machines were structured on a production line. Another large-scale project consisted of restructuring the production management system and computerizing it. Existing production management software packages were not however suitable for this type of industry, and a specific system had to be developed progressively. The first step, the preparation of production routings, was to serve as the core of production management onto which other functions were to be grafted.

2 Expertise in action: production planners prepare their routing

Before looking at a description of the expert system developed and the different stages of the project, we shall again consider the notion of a routing by means of a simplified example. This will enable us to detail the type of information contained in routings and to show the nature of reasoning behind their construction. Some difficulties associated with them will then be mentioned, allowing for a better understanding of the production planners' job.

2.1 Example of a production routing

If, for example, one wanted to produce 10 kg of an 11 mm-wide and 0,41 mm-thick strip in an alloy with a high silver content, the production routing given to the workshop would look something like the one in Figure 1. The first part of this document contains a fairly detailed description of the product and of the order, i.e. type of product (here a strip), dimensions, type of alloy, metallurgical characteristics, quantities ordered, quantities to be used, customer's name, order number and accepted lead time. The second part gives the sequence of operations. First it specifies the metal to be used as a base. In our example a 2,5 mm-thick ingot is taken from a stock of rough rolled ingots of differing thicknesses, in the most commonly-used alloys. (These stocks are built up to avoid the use of a whole ingot each time.) The first metallurgical operation consists of rolling the ingot to the required thickness. The routing specifies the machine to be used but also gives technically viable alternatives. When rolling is complete the metal is cleaned, before being cut into strips by means of a shearing machine; in this specific case five strips can be cut from the rolled sheet. The strips are then annealed to obtain the required mechanical characteristics. Finally, the order is checked before being packaged.

2.2 Preparation of a production routing: steps in the reasoning

How do production planners go about preparing a routing? At the start they are already familiar with the normal sequence of operations for manufacturing a strip: rolling, shearing, annealing. They also know that it will probably be necessary to roll the metal several times, on different mills, if the difference between the thickness of the first strip and that of the final product is too great. For this reason they will start with a 2,5 mm rough rolled ingot in stock (the semi-finished products mentioned above), and will reduce the thickness to 0,41 mm by rolling it on the same mill several times.

Having determined the thickness of the metal with which to start off, the production planners then have to determine its other dimensions. Rough rolled ingots are available in several standard sizes corresponding to the possible fractionation of an ingot reduced to that thickness. In each available size the planners calculate how many strips of the required width can be cut, bearing in mind the loss of metal on the sides, due to rolling, and between the sides, due to shearing. They then choose the width which will make it possible to minimize scrap and lost metal. In this case the width defined is 75 mm, which makes it possible to fit in five strips. They then have to determine the length of the rough rolled ingot to obtain 10 kg of strip, taking into account the scrap at each stage. In order to calculate this they apply a yield coefficient. The characteristics of the basic metal having been entirely specified, the planners must then describe the sequence of operations to be carried out to obtain the final product.

It is possible to reduce the thickness from 2,5 mm to 0,41 mm by rolling the ingot through the same mill several times. At this stage there is a variety of rolling

SILVER ROLLING	ALLOY 1 ARGELEC 2021	ALLOY 2	ALLOY 3	Department	Order N° 40000
Consignee : MARSEILLE				Quantity ordered : 10 kg	
Customer :				Delivery lead time : 8838	
Product description : **STRIP**		Plan : 11232 Spécification :		PLANNERS INITIALS	
Dimensions :	Length	Width 11	Thickness 0,41	STATE : NP HARDNESS : 75	
Tolerances :		+0,10/-0,10	+0,01/-0,01	Quantity scheduled :	

OPERATION		MACHINE	SECT	DIMENSIONS : DESIRED VALUES	TIME SPENT	ESTIM. QTY.
No	DESCRIPTION					
1	ROUGH ROLLED INGOT	SAMPLE		THICKNESS : 2,5 WIDTH : 75 LENGTH : 1000		
2	ROLLING	L.6.CYL		THICKNESS : 0,41 MILL ALLOWED 1. : DUO MILL MILL ALLOWED 3. : DUO AD8 MILL MAX THICKNESS TOLERANCE = +0,01 MIN THICKNESS TOLERANCE = -0,01		
3	CLEARING	WHITE 2262				
4	SHEARING	SHEARER 85		THICKNESS : 0,41 WIDTH : 11 RAISED NUMBER : 5 MAX WIDTH TOLERANCE : 0,10 MIN WIDTH TOLERANCE : 0,10		
5	FINAL ANNEALING	STOPPER 1500		THICKNESS : 0,41		
6	CHECK			DIMENSION MECHANICAL CHARACTERISTICS		
7	PACKAGING			WEIGHT		

Figure 1 – Example of a routing

mills which could be used. To avoid having to reintroduce the metal each time – which requires an additional operation and results in a loss of more metal – the planners propose a six-cylinder reversible mill, but also suggest two alternatives which do not have the same technical advantage. In this particular case they will naturally opt for a five-bladed shearing machine to simultaneously cut all the strips. The choice of a furnace will be determined by the size of the strip; since the metallurgical state has not been specified, it is not necessary to allow for particular processing conditions (e.g. controlled atmosphere). In each case the planners define the manufacturing constraints related to the product and compare them with the capacities of the most appropriate machines.

The above example has enabled us to familiarize ourselves with the notion of production routing and to better understand the reasoning behind it. We shall now examine several difficulties associated with the concept.

2.3 Production routing: a complex task

Several elements contribute to making the preparation and management of routings highly complex. We have already mentioned some of these, related to the actual activity of processing metals, i.e. the diversity of orders and the multitude of routing alternatives. But there are other factors which are specific to routings themselves. We shall examine these now, and identify, for each one, the choices made at the start of the project.

We have seen to what extent a routing is the result of a combination of variables. Its preparation involves a vast range of expertise which, in an established firm such as the one under consideration here, is often hardly formalized. It belongs to the production planners and workshop supervisors who acquire it gradually and empirically. The result is that there is no "optimum" routing as such. For each case several different documents could theoretically be drawn up, with the possibility of selecting the best one (depending on the meaning given to "best" – whether seen from an economic or quality point of view, for example). In practice, however, only one is prepared. We shall see that this choice has been incorporated in the expert system under consideration, which will not look for the optimum routing, but rather stop once it has found a satisfactory one that is compatible with most of the requirements[1].

A second difficulty lies in the division of roles, introduced by the routing and its contents, between planning and operation. There are two aspects to this question. On the one hand, how detailed should the dividing line be between technical and manufacturing expertise? Should operational modes or shop practice be described? Workshops have hitherto enjoyed a large degree of latitude and operators have always been left to set the machines, for example, or to determine the number of times that metal sheets are rolled. Instructions are only given rarely, to draw

---

1   Some attempts have however been made at having expert systems work out optimum routings for machining operations (Tsang, 1987).

the operator's attention to a specific requirement that differs from standard practices. On the other hand, who will make the decisions: the production planners or the supervisors? What decisions will each have to make? The choice of a machine well illustrates this problem. Production planners will identify what they consider to be the most suitable and efficient machine, depending on the characteristics of the order. But the supervisors may opt for a different machine, since they are directly responsible for executing as efficiently as possible the work schedule given to them, with the men and machines at their disposal and by taking into account all kinds of possible mishaps. They may therefore prefer using a different rolling machine even if it seems to be less efficient for that particular order, especially if their choice allows for a globally more efficient use of available resources. In principle, the expert system was at first meant to respect prevailing practices and not to change the balance between production planners and workshop supervisors in any fundamental way.

Even though it is difficult to define the notion of an "economic" routing, some significant financial choices are made implicitly during the preparation of these documents. The choice of the most efficient machine or the calculation of a maximum yield (discussed above) immediately come to mind as examples. There is a third aspect, that of the choice of the basic metal. Very often a mere fraction of a billet or ingot is sufficient for one order, in which case the same ingot will be used for several different orders (this type of production is said to be divergent). It is not however possible to divide the ingot from the start into butts corresponding to each order; it must first be broken down roughly. But even at this level, the quantities required for each order could be too small to be processed separately. Furthermore, such separation would decrease the yield (each time the metal is introduced into the rolling mill the initial strip is lost). In order to avoid this phenomenon the metal must undergo a maximum number of operations before the orders are distinguished from one another. A significant part of the production planners' job is the grouping of orders into batches. The amount of scrap metal generated and the number of operations required – two key variables in the performance of the workshop – depend largely on these grouping strategies.

Furthermore, long routings containing many operations result in increased lead time. In order to reduce this time, one solution is to start production at an intermediary stage rather than from an ingot. Several key thicknesses are determined, and semi-finished products stocked. For each order an attempt must be made, as far as possible, to start with a rough rolled ingot from these intermediary stocks, rather than from an entire ingot. This policy of semi-finished products is in keeping with that of constituting batches.

3 The reconstruction of automated knowledge: birth of a new type of routing

Rather than trying to develop an entirely specific expert system or using a generic system capable of solving anything from medical diagnosis to industrial scheduling problems, the firm preferred opting for a system dedicated to problems of managing technical data and preparing routings. It chose TOTEM (*Traitement Optimisé des Temps et des Matières*), developed by the company MWM. We shall first describe this tool, before explaining how it was adapted to the specific problem of production routing in the firm.

3.1 TOTEM: an expert system for generating routings

TOTEM is an expert system with relatively standard principles. The problem under consideration is fed into the system in the form of data or facts (here, a description of the order). An operation mechanism (the inference engine) then chooses and mobilizes in the knowledge base those facts which may be useful for solving the problem. The knowledge base groups together all the expertise required to prepare a production routing for all the orders it could possibly receive. This expertise is modelled according to rules which indicate that if certain conditions are met (that is, if certain facts are known to the system), new facts can be inferred. From a fact known at the start, the inference engine detects a rule that can be activated (the fact corresponds to the condition to be met) and infers a new fact through the application of the rule. This new fact will in turn make it possible to activate new rules and to infer other facts. The reasoning is thus propagated by a series of inferences until it is no longer possible to add to the known facts by applying new rules. The system then stops and the results produced are used in production documents, in this case routings. We shall add more detail to this description of TOTEM by identifying several characteristics of the system and its functioning which had a significant impact on the progress and contents of the project.

3.1.1 An empty shell endowed with knowledge-representation formalism

Like all expert system shells, TOTEM is basically an empty structure. Before the system can be used, the expertise and rules which will feed the knowledge base must be formalized and written. This formalization of knowledge must however be consistent with the structures allowed for in the expert system, i.e. the types of description of objects allowed by the system, the modes of representing knowledge offered by it, as well as its principles of organizing and structuring knowledge and its reasoning mechanisms.

Thus, the relations which exist between the elements dealt with in the manual preparation of a production routing must be expressed in the form of rules. Some of them will relate to the transformation of one of the characteristics of the product during a stage of the routing, for example the relation between the variation of thickness and the variation of the other dimensions during a rolling operation or the yield calculation. Others will direct the choice of operations or machines and will specify which setting will allow for a particular characteristic to be obtained. Although these rules model different kinds of knowledge, e.g. formulae for calculating parameters, choice of operations or of machines, etcetera, they use the same syntax: "if... then...". The "if..." part indicates the conditions which, if met, will allow the second part of the rule to be applied. If it is a matter of choosing a machine there will be a rule for each machine in the factory, whose conditions will define the possible use to which it can be put (range of thicknesses, maximum width, etc.). For the evaluation of a parameter, the condition part will simply indicate that the value of the parameter is unknown and the action part will provide the formula which makes it possible to calculate it.

Examples of the rule:
IF the metal must not be hard drawn after rolling
THEN it must be annealed
IF during rolling the thickness goes from $T1$ to $T2$
THEN the length of the sheet goes from $L1$ to $L2$ with $L2 = L1 \cdot T2 / T1$

This representation in the form of rules leads to an extremely fragmented formalization of knowledge, which corresponds fairly well to the way in which the production planners express themselves when questioned. Moreover, to favour the readability of rules and their applicability to numerous situations (in order to avoid partial rewriting), the project leader limited the number of conditions in each rule even if this meant multiplying the number of rules, and did not hesitate to introduce intermediary calculations into his formulae to make them shorter and more comprehensible. These choices, combined with the intrinsic abundance of knowledge and possible situations, makes it necessary to formalize an impressive number of rules. There are several thousand in this particular application.

3.1.2 Abundant expertise to structure

When the system was being developed it was clear that the multiplication of the number of rules was a potential threat to its efficiency. The system would have to scan the entire set of rules each time in order to choose the one it was to apply – something which, in view of the number of rules, could be excessively

3 The reconstruction of automated knowledge: birth of a new type of routing   141

time-consuming. To avoid this handicap the knowledge base was structured in detail. On a first level, different knowledge bases were distinguished according to families of products (knowledge varies greatly from one family to another). The knowledge on each product was then broken down into relatively autonomous areas of expertise, which would permit each one to deal with an aspect of the problem (e.g. an area of expertise on the rolling mill chosen, one on the rough rolled ingot chosen, or one on the processes required to obtain a specific metallurgical state). These different areas were not of course totally independent from one another and were organized into hierarchical structures, with easy access between them. Figure 2 gives the organization of a part of the areas of expertise for the strip family.

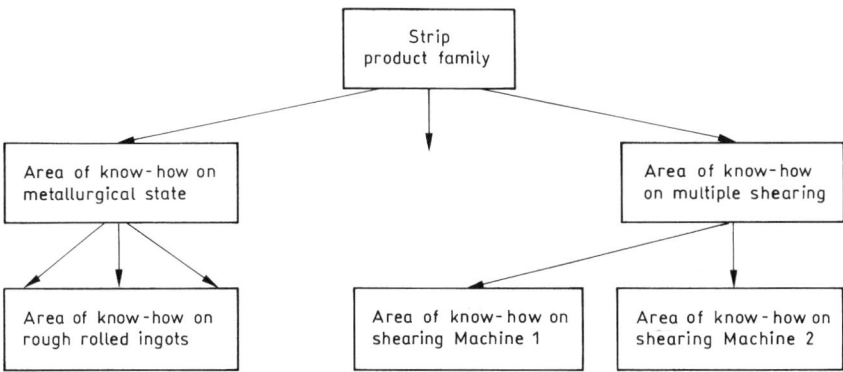

Figure 2 – Example of knowledge base structuring for the strip product family

The introduction of an order into the knowledge base led to a limitation, or more precisely a distortion, compared to the initial principles of expert systems where it was meant to be possible to write knowledge in bulk, in any order, and leave it up to the system to detect the applicable rules at each stage of its reasoning. In practice this principle is rarely applied, and knowledge bases are often split up so that at any moment the system only works on a relatively small number of rules, amongst which it is almost sure to find the one it needs, so making a larger search unnecessary. A second objective of this structuring of the knowledge base consists of facilitating the creation and maintenance of expertise; it is easier to identify a rule that is known to be in a certain area of expertise. Since this structure groups together a set of homogeneous expertise in a single area, it is therefore easier to check that all possible situations have been envisaged. Furthermore, the organization of these different areas suggests relatively clear macro reasoning, corresponding to the steps in the manual preparation of a routing, even if, as we shall see, the system does not really reason in the same way.

3.1.3 TOTEM's calculations: evaluating parameters

At this point it is important to note that the expert system TOTEM does not manipulate routings; it does not even build them, so to speak, but merely evaluates parameters by means of rules introduced into the different areas of expertise. These parameters are then matched to a standard routing for each product family, by a relatively standard database manager coupled to the expert system. Figure 3 represents the architecture of this system. The characteristics of each new order are entered into the database and added to its files of "articles" and "orders". This information serves to construct a TOTEM identifier, that is, the description of the problem which will be transmitted to TOTEM in the form of the list of principal characteristics. For a strip, for example, it will include the quantity to be produced, the specification, the plan number, the thickness, the state, the hardness, the packaging, the width and thickness tolerance, the minimum and maximum length if the product is delivered flat, and so forth. It also makes it possible to identify the family of products concerned and the corresponding area of expertise in TOTEM.

All orders received during a given period are transmitted in this form to TOTEM. It then uses rules to calculate all possible parameters and particularly those which describe the related operations, machines, dimensions and instructions. TOTEM stops when it can no longer evaluate any new parameters (it evaluates between 500 and 1,000 for each routing). The parameters needed for a routing are transmitted to the database manager and stored in the orders and operations files. The routing is finally prepared from the information contained in these files.

How, and in what order, does TOTEM calculate these parameters? This is done according to two objectives: first, that of refining the description of the product (for example the quantity to be started with is inferred from the quantity to be produced and the calculated yield), and secondly, that of selecting and describing operations with the relevant machines, dimensions, and instructions. To infer a new parameter, TOTEM uses a rule as well as parameters which have already been calculated. There are two types of parameter: explicit parameters (those of the problem) and implicit parameters (those inferred by TOTEM). Its main objective is to choose and to describe the operations to be performed, but to save time it will scan all implicit parameters and calculate those for which it has information, before attempting to choose operations. At the start it will not, for example, have all the elements required to determine the dimensions and instructions; it is then quicker for the system to first try to work on the facts it already has. When it has calculated all the possible parameters concerning the operations, it starts again and calculates new implicit parameters (if, for example, a rolling operation has been chosen, it will be able to evaluate the partial yield coefficient relative to this operation).

3 The reconstruction of automated knowledge: birth of a new type of routing

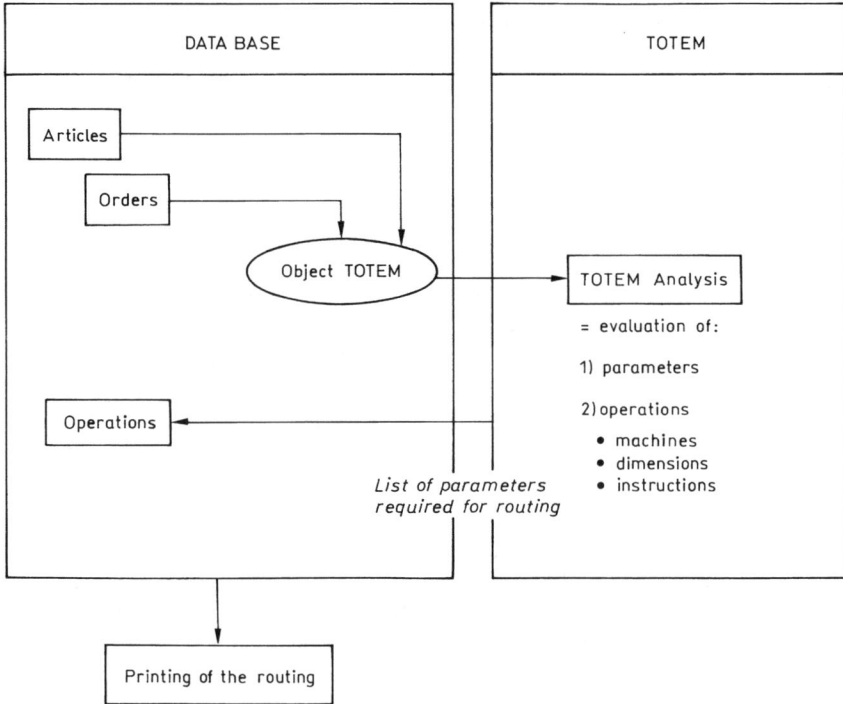

Figure 3 – Architecture of the system

3.2 Developing TOTEM: the procedure

We shall now present the main features of the development of the project, before looking at the essential step of formalization in order to analyse the nature of the problems encountered. We shall then describe the implementation of the system and its conditions of integration, and finally attempt a first evaluation of the project in its present state.

3.2.1 Nomination of a project leader with an extended mission

In 1987, when the firm decided to restructure production management entirely, it chose to create a new management post situated outside of the factory's hierarchy although linked to its management structure. The incumbent's mission was to draw up a coherent plan for the reorganization of production management and to implement the first stages, in particular the automatic generation of routings. The engineer appointed to this post was particularly well-suited to the job. He had managed a large metal processing workshop for several years and was therefore thoroughly familiar with the work, even if he still had to acquire some knowledge peculiar to precious metals. But most importantly, he already had experience in using TOTEM to prepare production routings. This was considered to be particularly valuable and was a decisive factor in the choice of a system.

3.2.2 A cautious and progressive development and diffusion strategy

The notion of areas of expertise and structured knowledge was used to promote a step by step approach. It was possible to start work on a single product family by progressively building up corresponding areas of expertise. This approach had the advantage of making it possible to obtain the first expert system routings rapidly. Although rudimentary, they gave substance to the tool, placing it in a favourable light during discussions about its development and use, and helping to make it better understood and accepted in the factory. They also provided concrete examples of the results which could be expected from it. The approach moreover helped users to familiarize themselves more rapidly with their future instrument by giving them an idea of the reasoning or at least the functioning of the system. A second phase of increasing and refining the knowledge bases could then be undertaken on this first product family, whilst the extension of a second family was initiated concurrently.

3.2.3 Early involvement of future users

Another characteristic feature of the project lay in the involvement of the production planners. Since the project leader's aim was to build a tool which would be useful to the planners in their daily work, he involved them actively in its development. Indeed, one of the recognized advantages of expert systems is that they provide experts themselves with the opportunity of formalizing their knowledge and of ending up with a "tailor made" tool perfectly suited to the questions they ask and the use they wish to make of it. The project leader's objective was based on his former experience, where production planners had contributed significantly to the formalization of knowledge and had progressively become so familiar with the system that they were able to take over the maintenance and updating of data within it. There were thus two main reasons for rapidly involving the relevant production planners in the project: on the one hand, it was inconceivable that the gathering, formalization and writing of information could be done without the firm's routing experts and, on the other hand, their active contribution to the development of the system was a guarantee that it would be suited to their requirements.

3 The reconstruction of automated knowledge: birth of a new type of routing   145

3.2.4 The first routings: the fruit of considerable effort

Modelling started with the strip product family in the rolling workshop. This option was guided by the relative technical simplicity of these products (e.g. well-defined and repetitive sequences, relatively clear choice of machines, and so forth). The relevant planners, whilst continuing their operational tasks, took part in numerous meetings with the project leader in order to clarify and explain the knowledge required to prepare "strip" routings, to formalize it in accordance with the logic of the system, and to write and organize this significant volume of information. This work also involved discussions with supervisors, in which the planners' expertise was compared to workshop practices and experience. Nine months of work went into endowing the system with a knowledge base large enough to produce its first routings. Certain corporate managers considered that this was excessive, but it could be explained, as we shall see below, by a number of problems – in some cases unexpected – encountered during knowledge formalization.

3.3 Knowledge formalization: an instructive phase

This phase warrants particular attention. On the one hand it is the main step in the process of developing an expert system, and on the other hand it was – in our case – the one which demanded the most time and energy to overcome the obstacles encountered.

3.3.1 Problems of formalization: transforming experience into laws

What explains why the knowledge formalization stage took longer than expected? The fact that the time-schedule was based on the project leader's similar experience in another workshop should normally have precluded such a discrepancy. However, several significant differences between the two industrial situations gradually became apparent and they help to answer our question. The first difference concerned the degree of preliminary knowledge formalization at the start of the project. In the firm we are studying, there were very few written documents specifying technical rules or describing procedures for preparing production routings. Most knowledge was informal, based on the production planners', technicians' and supervisors' experience. This absence of documents meant that an additional stage had to be introduced, in which their expertise was explained. In contrast, in the other firm knowledge formalization had been based on a large number of documents with detailed descriptions of the choice of operations, machines, operating methods, and so forth, which led to considerable time-savings. These documents owed their existence to quality guarantees provided by the firm to its main customers in the aeronautics industry. Such contracts provide for control of the production process by means of a formalized description of that process and of relations between the targeted results (mechanical or metallurgical characteristics) and the setting or choice of machines. We note in this connection that the considerable task of formalization bore the mark of numerous "methods" engineers, whose function was to improve production techniques and develop ways of processing new alloys. The firm under consideration did not have this type of resource and its design department was only responsibe for working on new orders. The question of creating a knowledge base was therefore posed in different terms in the two firms. In the one case it consisted of transferring a set of knowledge that was already modelled and available in an easily formalizable form. In the other case (the firm we studied) the entire task of explaining and modelling the knowledge had to be carried out before it could be fed into the system. The second element which differentiated the two projects was an effect of size. In the latter case there were only one or two production planners per workshop section, whereas in the other firm there were relatively large and structured teams of about 20 persons. The small number of personnel limited the technicians' availability with respect to the project.

Meetings between the project leader and the production planners soon brought to light the difference of approach between the latter and the expert system. In every new situation planners try to identify known problems for which there are clear and immediate solutions. In contrast, expert system reasoning is automatically broken down into elementary stages (fairly similar to those that we presented when introducing the notion of routing), each stage being formulated independently by means of a rule. The planners gave the impression of taking many shortcuts in their reasoning; their practical experience led them directly to the solution, without them having to examine the sequence of intermediary stages. Breaking down their reasoning into elementary stages was therefore totally unnatural for them. It was consequently difficult to determine the degree to which a reasoning process should be broken down into elementary rules and this stage necessitated much explanation and numerous discussions with the project leader.

The problem of identifying a reasoning process was compounded by the difficulty experienced by the production planners in translating their experience into general laws (expert system rules). It was no longer a matter of using their expertise to solve a given problem, but one of expressing *a priori* a body of knowledge which could be applied to all possible situations of a particular kind. Once again, this exercise in abstraction could only be carried out successfully with the constant involvement of the project leader.

Finally, the breaking down of reasoning into elementary stages, or of expertise into autonomous rules, only made sense with the articulation of the knowledge. That brings us back to the problem mentioned above. How many areas of expertise should there be and what is the relationship between them? Must each one of them answer a certain type of question? How can one ensure that all potentially useful

3 The reconstruction of automated knowledge: birth of a new type of routing

knowledge has been formalized? At the start there were no clear answers to such questions. An understanding of this arrangement of expertise into separate areas is essential for understanding the system's functioning.

Two standard questions which arise with the development of a knowledge base are related to completeness and coherence. Wondering about the completeness of the base amounts to wondering whether there are not gaps in the modelled knowledge, which could lead to an interruption of the reasoning process, or whether any unexpected situations could possibly arise. Aiming for coherence means ensuring that there are no rules with contradictory contents, nor rules which create loops. In all cases the difficulty lies in the need to take a global view of the knowledge base which has in fact been built up progressively by the quasi-independent addition of rules. Tools were designed to help users of TOTEM in their tasks of designing and developing areas of expertise. These consist mainly of the arborescent representation of sequences of rules, which explain all possible reasoning by the system "around" the use of a given rule. Thus, if an error[2] is identified in a routing produced by TOTEM, the results given by the system are invaluable for locating the origin of the error. However these tools generally consist of extensive listings which combine hundreds of rules, and are therefore not particularly easy to use.

3.4 Division between technical expertise and operational expertise

We have seen that modifying the balance and allocation of tasks between production planners and the workshop was not an initial objective. Thus for "strips", the first product family introduced into TOTEM, the expert system chooses the machine that is technically best suited to each operation, but also indicates other possible options so that the workshop can make the final choice. This choice is dictated by operational constraints peculiar to the workshop, which are not always known beforehand. TOTEM's routings are therefore fairly similar to those drawn up formerly by hand. They retain the same amount of detail in their descriptions, for instance the number and nature of steps mentioned or the operational modes and work-station instructions. The knowledge representation stage was nevertheless an opportunity for certain modifications and developments which we shall illustrate by way of the following examples:

– *The choice of thickness and other dimensions of rough rolled ingots*: We have seen the advantages of a strategy of stocking semi-finished products in this type

---

2 Error in the broad sense of the word: it may be that a step in the production procedure was left out, which makes the routing inapplicable, or simply that a discrepancy exists in relation to the routing that a planner would have prepared himself.

of industry where production routings contain many steps. The expert system will therefore try to start with a rough rolled ingot whenever possible. This policy was already applied before the introduction of the system, but was systematized and rationalized through its implementation. The determination and management of these stocks of rough rolled ingots was not, however, straightforward. In what form and in what quantities were they to be stocked? Knowledge representation provided an opportunity for specifying the most suitable thicknesses for the ingots. (Thicknesses are inferred from the characteristics of the available machines and orders, and dimensions attempt to optimize the fractioning of an ingot or a billet reduced to a given thickness.) The number of intermediary thicknesses was reduced and physical stocks were rationalized by geographical grouping of stocks and control of available stock levels.

– *The updating of work station instructions*: A production routing does not in general describe operational modes for machines, except where a specific procedure must be applied which differs from habitual workshop practice. In that case an instruction is given on the routing. During knowledge formalization all such instructions were re-examined with the production planners and workshop supervisors and an updated list specifying their conditions of application was drawn up. Some of them were modified, or even eliminated, so that the list accounted for the effective practices and expertise of the workshop.

– *The definition of certain concepts*: This applied to tolerance, for example. In production routings, dimensions are generally accompanied by their tolerance (+ or − so many millimetres). The planners had become used to noting the greatest value of the requirement, but what meaning was then to be given to a tolerance which was not a requirement? For the sake of clarity, only the notion of tolerance was retained in TOTEM, indicating the range of compulsory parameters.

Work carried out on other more complex product families showed, however, that is was not always possible to draw a line between planning and production, and to safeguard the workshop's autonomy in taking operational decisions. Two examples demonstrate how the objective of optimization could lead to certain choices, hitherto made in the workshop, being made during the planning stages.

The first example concerns yield optimization for the sheet product family. We have seen that a strip is characterized by its thickness and width. To define a sheet, its length must also be specified. With the addition of a dimension the yield calculation is no longer made on a section, but on a volume. Generally an order is composed of several identical sheets. The problem then consists of finding the best arrangement for these sheets in a given thickness and of defining the operational sequence to maximize yield. This is a well-known problem in operational research, which uses linear programming in whole numbers. But the question is in fact even more complex here, since an ingot can generally not be kept whole when it is processed – it has to be broken up (because of maximum dimension

3 The reconstruction of automated knowledge: birth of a new type of routing   149

constraints in annealing furnaces, for example). The way in which this is done will of course have an effect on the yield calculation, since it will condition the succession of ulterior dimensions of each of these sheets during their respective processing. One can therefore understand that yield optimization, associated with the production of these sheets, leads to the definition of a specific operational sequence. The choice of a basic metal which conforms to the dimensions given by the routing, as well as the choice of a sequence of operations, become essential requirements and failure to comply wirh them could make it impossible to obtain the required number of sheets in the order. The latitude enjoyed by the workshop is thus reduced.

The second example relates to the choice of a machine. Traditionally one of the difficulties in wire drawing lay in the choice of a machine and in that of a series of successive reductions of the diameters of the wire draw die. The number of possible combinations was indeed huge, with on the one hand 18 relatively versatile machines and on the other hand multiple series of possible diameters. The formalization of the areas of know-how provided an opportunity to clarify the choice of machines by specifying underlying technical stakes. A study carried out by the project leader and workshop supervisors identified the series of wire draw dies that were technically the most suitable. It was thereby possible to specialize the machines and so avoid tool changes which normally necessitated stopping the machine between two orders. In the end the combinatorics were reduced considerably; an algorithm entered into TOTEM now allows it to choose the most suitable succession of diameter reductions and thereby the appropriate machine. This example also shows the limits of a passive attitude vis-à-vis knowledge, whereby existing knowledge is merely transferred within the formalized framework of the expert system, without its validity and relevance being questioned. Our example is, in contrast, one of an effective production of new knowledge, the improvement of technical knowledge and the reorganization and rationalization of work in the workshop.

3.5 A judicious criterion for choosing applications: the existence of generalizable expertise

Another lesson was learned from the development of the first knowledge bases when a criterion for identifying those product families which lend themselves to the use of TOTEM, was made apparent. TOTEM can only be applied if the expertise used to prepare a routing is generalizable, that is, if it can be written in the form of laws which can be applied to a relatively large number of situations. This property depends on the technologies implemented and on the nature of orders handled. A first examination of the knowledge applied in the platinum workshop, for example, showed considerable heterogeneity in the products manufactured and the technologies and machines used. An activity is either very specific, and then requires special machines (for manufacturing dies for the glass industry, for example), or else it remains a craft, in which case the workshop artisan's expertise remains decisive.

4 The project and the development of stakes: towards an industrial transition

4.1 Implementation and conditions for the integration of the system

While TOTEM was being developed on a product family, computer terminals were installed in the offices of the production planners concerned. They started using the expert system in their work as soon as its routings were considered precise and reliable enough.

4.1.1 Use of the expert system and role of the planners

The production planners' work has been changed profoundly by the introduction of TOTEM. They now no longer create, but merely interpret, or sometimes complete, routings, and feed data relative to new orders into the system. The next day they receive the new routings generated by the system, approve them, and transmit them to the workshop. TOTEM also produces inspection tickets and progress sheets, which helps to facilitate administration.

Some essential aspects of the planners' work have not, however, been integrated into TOTEM and still require their direct intervention. These are mainly the choice of a basic metal, the constitution of batches, and the grouping of orders. In order to function, TOTEM needs to be told what metal will be used as a base, and what its characteristics are. It is up to the planner to make this choice. He will start with a rough rolled ingot, make sure it is in stock, and specify its dimensions. The work carried out during the development of the expert system to streamline the management of rough rolled ingots facilitates this task, e.g. redefinition of thicknesses and dimensions, as well as storage space, especially in the silver workshop. Because of a space problem, stocks of rough rolled ingots were formerly kept in different places in the workshop and often mixed with other stock and unfinished work. Consequently, they were not easy to find and only the supervisors could identify them. With the development of the expert system, stocks were reorganized and partly grouped together to be more identifiable. The creation of batches and grouping of orders now constitutes a preliminary step in the selection of a basic metal, since the intention is to take advantage of the characteristics which several orders have in common, so as to handle them together and thereby limit the number of operations in the workshop.

Orders do not need to be identical for this strategy to be applied; it is enough if they have common stages in the manufacturing process. For example, it is possible to take the same 2,5 mm rough rolled ingot to produce one 0,5 mm and another 0,2 mm strip in the same alloy, if their respective widths are compatible. Or, to optimize the use of an ingot, it is advisable to find a combination for which the weight, yield calculation included, is close to that of the base ingot. These strategies of grouping and batching are by no means simple and require sound experience in metallurgical techniques. They do, however, predetermine to a large degree the workshop's efficiency.

Finally, we note that new tasks can now be given to the production planners, such as more accurate control of operating time (e.g. refining the standards of time used to draw up quotations) or the control and attempted improvement of yield. But their mission is broader than that, for they must also contribute to the maintenance and upgrading of TOTEM's expertise. In this respect, the situation varies according to the product family and the production planner. Some are proficient enough to be relatively independent when it comes to updating the computer system, whereas others still need help in changing the rules or adding to areas of expertise.

4.1.2 Emergence of a new actor: the methods and planning engineer

Although nobody imagined that the system could be developed without relying heavily on the production planners' expertise, it was similarly inconceivable that they should carry out this task alone. The project leader played a key role at each stage of the project, be it in explaining expertise, in translating experience into general laws, in teaching the syntax for writing rules and the mode of operating the computer, or in research carried out with the managers and aimed at improving expertise before formalizing it; not to mention all the problems of setting up the computer system, nor negotiations to modify or adapt the software and its contents. The project leader saw this mission as a temporary one, lasting roughly the time required to develop the tool. The planners were meant to acquire enough autonomy during the course of the project to run and upgrade the system themselves. We have seen that this objective was not entirely achieved.

In these conditions, a person capable of fulfilling a permanent role of assisting the planners, had to be found. In order to define the most suitable profile, careful thought had to be given to the exact mission that such a post should have, particularly since the project leader was in the meantime appointed as operations manager of one the departments in the factory. An engineer who had formerly been responsible for production management was selected. He did not, like the project leader, have the mission of designing a new system of production management nor of setting it up; his function was rather that of a methods engineer, for production itself and for assisting the planners. Whilst this function was normally associated with production problems, it was new for planning. In fact production planning is, in relation to production itself, in a position similar to that of methods since it provides it with a guide. With the formalization through the expert system of the intellectual tasks accomplished by production planners, the necessity of proposing assistance and rationalization for these tasks – similar to that enjoyed by the operators of workshop machines, – became apparent. These two types of task are now described.

4.2 First elements in the evaluation of a project

Three years after the start of the project, it is possible to make an early evaluation of it.

4.2.1 The present situation

TOTEM will finally be used for routing in three workshops only: silver rolling, silver wire drawing, and machine-finished gold, while more conventional methods will be used in the other workshops. It already prepares routings daily for the strip product family in the silver rolling workshop and three thousand rules have been written to formalize the corresponding expertise. Knowledge bases for metal sheets and wire drawing have been developed, the routings for these product families are being validated, and the two applications should soon be used operationally by the production planners concerned. TOTEM's routings are judged satisfactory by both the planners and the workshop; they are fairly similar to those that were formerly drawn up by hand, although more complete.

We note, however, a result which gave rise to debate. With certain product combinations it has been possible to find, by very careful calculation, a better yield than that proposed by TOTEM. For example, the latter planned the use of 4 ingots when 3 would have been sufficient. Such situations are fairly rare but they do show that there is room for improving the local optimization of the system.

It is furthermore still a little early to judge the impact of these routings in the workshops. For example, will changes in instructions influence behavioural patterns? Will the conditions for using annealing furnaces be respected? Questions like these necessitate a comparison with workshop practices to validate the expert system's instructions, improve them where they are unsuitable, or explain their meaning and advantages to the operators.

The decision to proceed progressively and cautiously has proved to be a good one. It has made it possible to overcome formalization problems whilst gradually informing everyone in the firm about the new tool. Such problems, which led to delays in the progress of the project, can be explained by the absence of preliminary formalization of the planners' empirical know-how, by the abstract nature of modelling and by the volume of know-how. Furthermore, it was perhaps a slight mistake not explaining TOTEM's reasoning more thoroughly to the planners. It would certainly have been better to look beyond a mere presentation of the rule syntax (if... then...), at the structuring of knowledge into specific areas and the use made of these by the system. These difficulties raised a question during the course of the project: who was going to manage TOTEM's knowledge base? In the end an alternative solution had to be found, since the production planners were not yet fully autonomous with respect to the system. We have seen how the idea of a methods engineer gradually took shape. This evolution was not surprising, for the expert system can be likened to a machine, it is a technical system which has to be maintained and managed as such.

4.2.2 The role of economy in the project

From the start of the project, specifically economic considerations did not play a decisive role. The main idea was not to achieve immediate productivity gains by reducing, for example, the number of production planners, especially since there was generally only one planner per workshop, except for rolling, where there were two. Rationalization for TOTEM should rather be sought in the formulation of expertise and the facility to make it evolve. For example, the new properties of alloys can now be itemized directly; or if the cold-rolling curve is to be moved only a single point need be changed to obtain the new position for annealing the metal. Finally, a major objective was to build a first link onto which other production management modules could be joined.

4.2.3 The significance of indirect spin-off

A project like this cannot, however, be evaluated solely in terms of the intitial objectives assigned to it. There are frequently numerous unexpected results which can be exploited. The TOTEM project was enhanced by such results, for example: the improved definition of rough rolled ingots and of their dimensional characteristics, coupled with a reorganization of storage space allowing for a more systematic use of semi-finished products; the optimization of yield for sheets or for wire-drawing machines; or the introduction of new concepts which pave the way for restructuring and reorganization. Thus the notion of "filiation node" (a key step in the production routing) is used to define the steps in the processing of orders.

4.3 Emergence of a new type of methods engineer

The main feature of the TOTEM project and the most positive element in its development lies in the fact that it led to the emergence of a new role, that of the methods engineer. The evolution of the variety and complexity of products had led to a large body of expertise which needed to be restructured and ordered. In the wake of the automation of routings it was in fact the place and the role of production planning that was re-examined. Finally, we can characterize this project by the term "restructuring automation", whereby we mean that the project overtook the mere automation of existing tasks. This automation was only possible to the extent that it was accompanied by a profound reconception of the activity of production planning. Certain concepts were specified, others introduced, and finally the general rules which apply to broad types of situations were formalized.

Chapter 2
Cornélius
Fragmented expertise of maintenance specialists

The Cornélius case history takes place in an advanced mechanical production firm that manufactures parts for complex units subject to stringent conditions of utilization. Technological developments in this sector are particularly rapid and affect the materials used and the shape of parts, as well as production methods. Quality control has to be tightened as requirements are multiplied and international competition intensifies. To keep up with these changes and meet its goals, the firm recently undertook a vast automation programme. New technologies (robots, numerically controlled machines, flexible cells, CAD, and so forth) were introduced and have profoundly modified production modes. Maintaining this complex equipment has at the same time become a critical factor in the overall performance of the firm. In this context, the first expert systems devoted specifically to maintenance seemed to be worth investigating. An expert system development project was therefore launched on a specific production facility. This chapter presents the ES project's history.

1 Problem and context: controlling a facility

The factory in question has a service for promoting automation and modernization, and providing methodological data on the management of this type of project. Information is obtained from the results of pilot operations in which the service is actively involved. One of its main concerns is to keep a watch on all technological innovations and to attempt to evaluate them. Hence, it immediately became interested in expert systems for maintenance support and decided to embark on the development of an application. The aim was to familiarize itself with this technology and to assess its suitability for use in breakdown diagnosis and maintenance. Particular attention was to be paid to problems encountered during the development and implementation of the tool.

The first question which arose was that of selecting a facility for running the experiment. Several conflicting requirements had to be met: on the one hand it was necessary to choose equipment which was prone to frequent failures, so that the reasons for developing a diagnostic tool would be valid enough to motivate the relevant actors; on the other hand, it was preferable to opt for a technical system with limited complexity so as to limit the time spent on developing the technology and to evaluate it as early as possible. *A priori* the machining workshop with its numerous semi-automatic machines matched these requirements fairly well. The choice finally fell on the largest flexible cell in the workshop, which consisted of two machining centres equipped with NC (numerically controlled) machines, tool magazines, and an automatic supply, flanging and stocking system (see Figure 1; the facility and its functioning will be described in more detail in section 2.1).

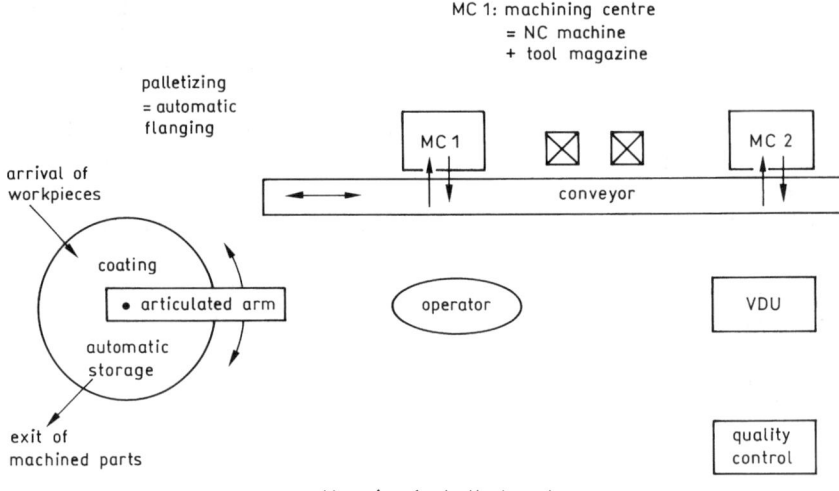

Figure 1 – Machining flexible cell

There were three main reasons for this choice. First, the flexible cell – essential in the manufacture of three different types of part – was a bottleneck in the workshop's production flow. Failures were relatively frequent and utilization was at about 80%, something which, in view of the volume of production programmed, posed a number of problems. Secondly, its complexity made it fairly representative of the problems which might be experienced with workshop machines. If the expert system was capable of diagnosing all failures on this flexible cell, it would also be able to identify the origin of failures anywhere else in the workshop. The cell integrated and managed automatically a significant number of systems and functions such as NC machining, automatic supply and stocking, and tool magazine management. (Later generations of cells used in the workshop had a simpler design.) Finally, it was thought that the presence of particularly competent operators and maintenance personnel would undoubtedly contribute favourably towards the success of the project and the later use of the expert system by operators accustomed to managing sophisticated computer systems.

2 Expertise in action: the maintenance team

The facility chosen therefore seemed to combine all the requirements for success. The next step consisted of selecting the team which would design the system. We must however first describe in more detail the functioning of this mysterious facility and the organization of production, particularly the attribution of roles between production as such and maintenance services.

2.1 Operators of a highly automated flexible cell

Seeing this machine in operation, one is struck by the fact that there is only one operator working on it, in spite of its considerable size and the multiple actions which seem to take place simultaneously. The man and the machine are occupied by different tasks and appear to be fairly autonomous, with each one acting independently. This is a far cry from traditional tool machines (e.g. lathes or cutters) where the operator was always seen standing over his machine, very busy guiding the different movements of the part or the tool holder, checking the size of the chips and the dimensions of the part on which he was working. It is at first not easy to understand exactly what the flexible cell is doing, for the sequence of operations does not appear to be completely logical, and especially because most productive operations are hidden from view by an enclosure which isolates the NC machines. The operator himself hardly touches the machine, but often consults a computer screen which informs him on its state. He can, by means of a keyboard, give instructions to the computer which will transmit them to the different elements of the machine.

How does this flexible cell function? When they arrive, unworked parts are first coated in an alloy gangue with a low melting point so as to facilitate their prehension and to ensure correct geometric positioning during processing. They are then stocked automatically in a magazine, while the computer saves data on their characteristics and position. When the production of a part is launched by the operator, the computer activates an articulated arm which takes a workpiece of the specified type from the magazine and positions it on an automatic flanging stand.

Once the workpiece has been secured, it moves with its stand on a conveyer that supplies the two machining centres. It stops in front of the door leading onto one of them and waits until the operation on the preceding workpiece has been completed. The door finally opens, the finished part leaves, the unworked part enters the machining centre, and together with its stand is fitted securely onto the NC machine. Machining operations can begin as soon as the door has closed. These take place in several stages, during which it is sometimes necessary to change the cutter or to modify the position of the workpiece in order to machine every side of it. All these operations are automatic. The NC machine executes a programme which specifies the nature of operations, their sequence and the tools that must be used. When the operations are complete the workpiece leaves the machining centre and is replaced by the following one. It then returns to its departure point, is removed from its stand and stocked in the magazine. To prevent the machining centres from remaining inactive, the computer initiates the preparation of several workpieces which will pass the finished ones on the conveyor. The choreography of what resembles a dance is indeed impressive!

What role remains for operators in such highly automated functioning? They still carry out a few directly operational tasks such as removing the finished parts from the remaining alloy gangue or doing some clipping, but their main role is elsewhere. It consists of giving orders to the flexible cell, of developing the production schedule according to the needs of workshops downstream and of supervising production progress. They also check the wear on tools kept in the machining centres' magazines and decide when to replace them. They control the quality of finished parts, for example the aspect of machined surfaces, or conformity with geometric specifications. Finally, they have to deal with all the breakdowns which may, and do, arise. In such events they first try to find a solution alone so that production may be resumed as quickly as possible, or at least so that the machining centres are still supplied. In order to do so they may have to prepare the workpieces themselves on a manual flanging stand, if the breakdown concerns the magazine feed attachment or the automatic flanging stand. If they are unable to solve the problem alone they call the maintenance service.

Before describing the organization of maintenance, it must be noted that this flexible cell was designed and introduced four or five years ago. It is one of the most complex and ambitious ever realized by the firm and functions continuously, without the presence of an operator at night. During the day only one operator is present, although he is assisted in the mornings by a second operator who deals with minor incidents which may have occured in the night, and performs maintenance tasks. These operators were chosen at the start for their skills. They report directly to the workshop supervisors and work according to monthly production targets.

2.2 A two-tier maintenance service

About a hundred employees work in the maintenance service, with on the one hand the workshop technicians or maintenance fitters, grouped according to their trade (mechanics, electricians, automaticians, etc.) and responsible for a set of machines in a particular operation area; and on the other hand a pool of about ten highly skilled technicians who take delivery of new machines from the supplier and supervise their use for the first year. These machines are then left in the hands of the workshop technicians who take over equipment that has been run in and of which the dysfunctions are well known.

In the case of the flexible cell under consideration, the complexity of the machine and the difficulties encountered led to a somewhat different approach being adopted. Although the cell had been in use for over three years at the start of the project, the specialist from the central pool was still working on it. He had, during this time, gained experience which would prove to be invaluable for the project. Nevertheless, this situation also was inherently risky since the specialist was the only person to have in-depth knowledge of the flexible cell. Moreover, the arrival of new equipment was soon to take him off this job and the problem of transferring his knowledge to the workshop maintenance fitters became acute. With this in mind, they had begun working on the cell.

The maintenance of the facility was therefore the responsibility of several actors belonging to different services, who had neither the same knowledge nor necessarily the same objectives with respect to the operation of the flexible cell. During the initial period, however, it was mainly the specialist from the central maintenance pool who would be concerned. He was undoubtedly the person in the factory most familiar with the functioning of the facility and was moreover used to systematic reasoning, well suited to the logic of the expert system.

3 Reconstruction of automated knowledge: expertise and usage

The active phase of the project began in 1988 and lasted about six months. Several more or less chronologically ordered stages can be distinguished. At first the project was focused on the choice of an expert system and on the adaptation of the techniques it employed. It then had to be applied to the flexible cell by modelling the latter to suit the system's logic. Finally, with the first experiments the question of its usage and the identity of its real users was raised. This itinerary could be described as an attempt to progressively contextualize the expert system. We shall now look at it more closely, step by step.

3.1 Cornélius: an expert system dedicated to maintenance

A number of experiments with mediocre results confirmed that early expert systems which only used knowledge expressed in the form of "if A then B" production rules, were ill-suited to the development of diagnostic applications. With such systems it was only possible to describe a list of symptoms and to link them to their origins, but that type of approach was far too limited for diagnosis where it was necessary to model the internal functioning of the machine and in particular the relations between its basic components. The designers of expert system shells[1] soon undertook to do just that.

When the project was launched there were still relatively few expert systems devoted to industrial diagnosis and maintenance, with only two suitable software packages: Maintex and Cornélius. The latter system was finally chosen, mainly because its basic concepts were simpler. The project leaders thought that this would simplify the training of future experts and users and facilitate usage, even by the uninitiated in computer technology. As we shall see in more detail in the next paragraph, Cornélius adopts an operational approach in its reasoning; it only handles objects which can be controlled by the user (either to test their functioning, or to replace them and ensure that the system is running smoothly again). This type of reasoning has the advantage of resembling as closely as possible the reasoning of a technician with theoretical training.

Cornélius' functioning is not particularly easy to understand and the specialist of the central maintenance pool received a week's training by the software supplier to familiarize himself with the management of the system. Takeover of the expert system went off without any major problems and the technician found himself quite at home with the system. He found that it reasoned fairly similarly to the way he did when diagnosing failures, that is, by way of a rigorous step by step approach with very few short-cuts. He moreover liked having to ask users as few questions as possible.

How does Cornélius go about diagnosing a machine failure? We shall now present its functioning very simply, taking the flexible cell as an example. This facility is divided into its main systems and functions: automatic flanging, opening and closing of the machining centre's door, movement of the conveyor, change of tools, and so forth. Failures are manifested by symptoms which lead the system to suspect one of the possible causes. In order to do so it uses production rules such as: if "the door does not open" then "suspect the open door" function. The diagnosis does not end there, for there may be multiple reasons for the door not opening. The actuator which moves the door may be jammed, the electrovalve or its electricity supply may be faulty, or else it may be that the control digit has not

---

[1] This is the name given to a computer programme which provides a mode of reasoning and of representing knowledge, but to which a knowledge base must be added to obtain an operational expert system.

3 Reconstruction of automated knowledge: expertise and usage   161

reached the electrovalve and that the door does not know it has to open. But it is also possible that the door was in fact opened and that the receipt digit did not reach the master computer. In other words, a whole series of elementary operations are realized in the opening of a door; if just one goes wrong it causes failure.

How can the faulty element be identified? Each system and function can be described by a diagram, similar to an electric circuit, where the components are arranged according to their technical relations. Figure 2 represents this diagram for the function of opening a door. It is read in the following way: a digit controlling the opening of the machining centre's door is sent out by the NC machine. It is received by a triac plate which acts as a contractor and stabilized supply for electrovalves. This triac plate is equipped with a pilot light which makes it possible to check that it is on, and is protected from overvoltage by a fuse. The triac plate translates the entry signal and sends the electrovalve a current which activates it during the opening of the door. The electrovalve activates the actuator which opens the door. Finally, a signal is transmitted to the NC machine to inform it that the operation has been completed correctly. When the expert system suspects a major system or function, it follows this type of circuit step by step, making sure that the component at each stage functions properly and is not the cause of the failure. Although it does not in fact reason exactly according to these circuits, it does rely on maps of normal and deviating machine behaviour (see Figure 2). These consist of nothing more or less than a logical translation of the above circuits. The following example explains the interpretation of these maps of behaviour:

For example, the following extract of a map of deviating machine behaviour:

Actuator (deviating) $\rightarrow$ Reception (deviating)

can be read as:

1. The "deviating" state of the receipt digit can be accounted for by the "deviating" state of the actuator. The system then checks whether the actuator is "normal" or "deviating". If it is "deviating", it follows the sequence of dependence until it finds the faulty element, otherwise, that is if the state of the actuator is "normal", it concludes that the receipt digit is the cause of the failure;

or, more naturally:

2. If the actuator is "deviating" then the receipt digit is "deviating".

Similarly, for normal behaviour:

Actuator N $\rightarrow$ receipt digit N

If the receipt digit is "normal", it can be inferred that the actuator is also "normal".

The procedure followed by the expert system consists of scanning the maps and generating knowledge on the state of the components, either by logical inference or by asking the user questions. For each component the system has a list of tests which it can ask the user to carry out. The questions put to the user describe the procedures of tests which the system asks him or her to carry out, for example checking that there is current, changing a fuse, and so forth. The answers provid-

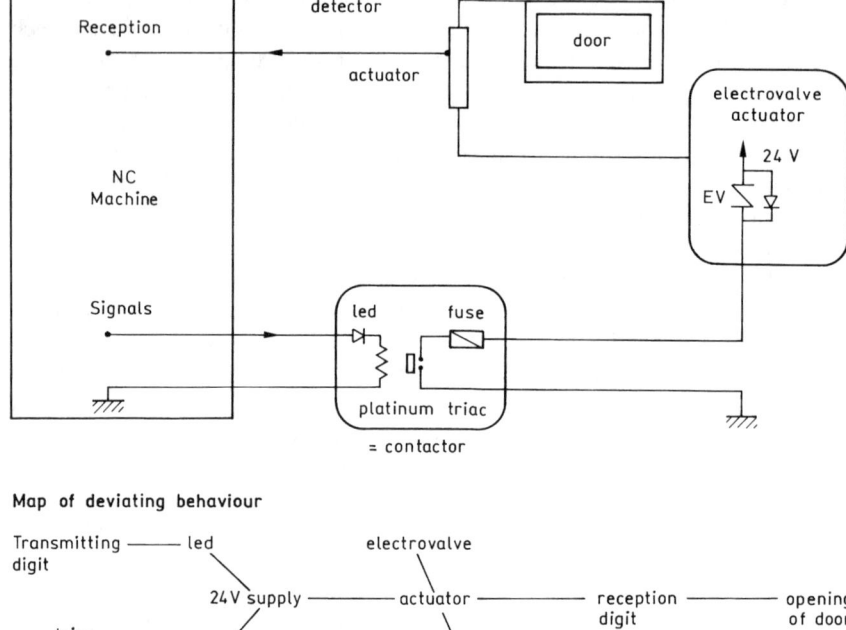

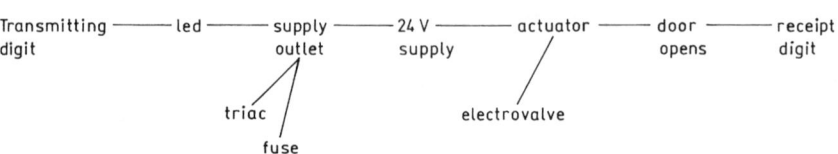

Figure 2 – The "door opening" function and its behaviour maps

ed by the user allow the system to infer the normal or deviating state of the component and to continue its search either by logical inference or by putting new questions to the user.

If at the end the system can still not detect a faulty component responsible for the failure of the function, it suspects a different one by way of the relations of dependence between functions.

3.2 Physiology of the machine: structuring knowledge bases

The expert system's reasoning is thus based on the principle that it should look at the facility's functions. The first step in its diagnosis consists of identifying the function which is most likely to be concerned by the failure and then to start searching from there. An advantage of this approach is that it makes it possible to structure the knowledge base (there is one knowledge base per function). Besides technically upgrading the expert system by way of the simultaneous manipulation of a limited volume of knowledge, the approach offers the possibility of progressively building up the knowledge bases, function by function, and also of learning from experience gained during the constitution of the first bases. Finally, the updating of the knowledge bases can be facilitated by this structuring.

Even before writing the first rule, it was necessary to choose sub-divisions. In fact the facility's control visual display unit, or VDU (i.e. the screen and keyboard which permit the man-machine dialogue) already proposed a first diagnosis; it indicated the origin of the dysfunction by comparing entry and exit signals with the theoretical values which they would have had if the functioning had been satisfactory (the signal is the command sent by the computer to the machine to perform an act, and the exit signal is the message sent back by the machine to show the computer the result of that act). If a discrepancy was shown to exist the VDU indicated the suspected function, with function taken here to mean a circuit comprising both a physical system and its control system. The operators knew how to interpret the messages which appeared on the VDU but, as we have seen, knowing which function is suspected is generally not enough to identify the faulty component. It therefore seemed logical to use this initial information to direct the diagnosis towards the suspected function, since the sub-division into functions adopted in the expert system was modelled on that already used for piloting the facility. This sub-division into functions was only useful in so far as the functions thus defined were relatively independent of one another.

Once the sub-divisions had been defined the specialist from the central maintenance pool could begin writing the first knowledge bases. The three functions most prone to failure were chosen first. These were: palletizing (or automatic flanging), management of the tool holder and tool change magazine, and the feedback control of speed and of the NC machine's position. The amount of modelling work seemed huge; it consisted of nothing less than a complete discription of the physiology and pathology of the flexible cell. In practice, however, the work was facilitated because it could be based on existing formalized expertise. Shortly before the start of the expert system project a maintenance manual for the flexible cell had been prepared by a work group. This group was made up of a workshop supervisor, two artisans who worked on the facility, a representative of the maintenance service and a representative of the service responsible for promoting innovation. It was to meet once a week for a year and prepare a detailed description of the facility and its components, the most frequent failures, the diagnostic procedures to implement and emergency repairs. This maintenance manual was seldom used by operators when failures occurred but it was useful in training new operators. The writing of the manual gave rise to numerous debates on the naming of the facility's components. The main idea was to remove all ambiguity which is generally prevalent even in technical language, and to agree on clear and distinct definitions. The wording of test procedures had similarly been a subject of much discussion and was left unchanged. Thus, a lot of time was saved in knowledge formalization, which only took two months.

The modelling of relations of normal and deviating behaviour of components of a function turned out to be the trickiest part, and the one in which the expertise was concentrated. Contrary to what one may imagine, this type of task does not consist of a mere translation, in a modified form, of functional plans prepared by the supplier when the machine is designed. The elements which will prove to be relevant for diagnosis must be extracted from this exhaustive representation. Thus the choices made here will be a determining factor during diagnosis, particularly on the number and order of questions put to the user by the expert system. Two types of expertise are required at this level. On the one hand, it is necessary to have sound knowledge of the process and desireable steps of a diagnosis (what questions must or can be asked and in what order, what tests can be carried out, and so forth); on the other hand, it is also necessary to be thoroughly familiar with the expert system's mode of reasoning, to anticipate the effect of a particular type of modelling of the functioning. Only this combination of expertise makes it possible to manipulate the system to direct the diagnostic sessions by adapting them to the user.

3.3 Experimentation and fine-tuning: who were the users to be?

Tests carried out with a first version of the knowledge bases did not entirely satisfy the expert system designers. The difficulties encountered can be illustrated by two significant examples. One concerns the repair strategy and the order in which questions can reasonably be put to a user, whilst the other relates to the limits of resources and knowledge deployed by users during diagnosis.

In the first one, the designers considered that the number of questions put to users was excessive and that the order in which they were asked was still not relevant. Their improved understanding of the expert system's reasoning led them to rewrite more efficient knowledge bases. They reduced the number of components described (certain tests on similar components produce redundant information) and rewrote the maps of relations between machine behaviour so that the series of questions would be more adapted to a diagnostic procedure. This point can be illustrated by the example of the contactor which we encountered in the "door opening" function (see Figure 3).

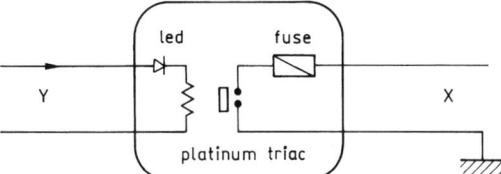

Figure 3

The map of deviating behaviour was initially as follows:

Y-D → Led D → Triac D → Fuse D → Supply D → XD

This modelling corresponds to the approach of a designer who might describe the functioning of the facility by following the power or information circuit. To explain the deviating behaviour of component Y the expert system will pose the following list of questions, in this order, until it has identified a component with normal behaviour:

- Are there 24V at the supply outlet?
- Is the triac fuse burnt out?
- Does changing the triac make it possible to start up again?
- Is the triac led on?

In order to answer the questions asked, users must perform tests which are sometimes fairly complex. For example, it is possible to see at a glance whether the triac light is on, whereas special equipment is required to measure the current at the supply outlet; similarly, changing the triac is time-consuming. The best diagnostic approach would consist rather of checking if the led is on before undertaking more difficult tests. An economic approach in the search for information may be more important for developing a diagnostic strategy than respecting the logical sequence of elementary components.

The following map illustrates the approach.

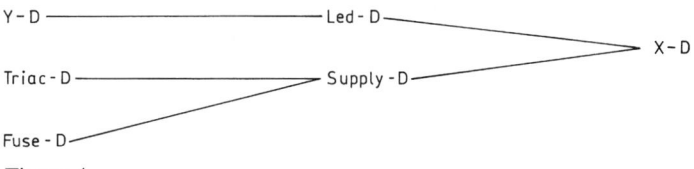

Figure 4

The order in which the questions are put to the user is then:
1) Is the triac led on?
If yes go to 2) if not perform test on Y.
2) Are there 24V at the supply outlet?
If yes X is faulty if not go to 3).
3) Is the triac fuse burnt out? If yes go to 4).
4) Does changing the triac make it possible to start up again?

This time the order of the questions will seem more natural to the user and the system will first ask him for information that is easy to obtain.

The second example is related to the attempt to extend the diagnostic procedure to breakdowns concerning the movements of the tool machines' axes, one of the main sources of machine failure. The system designers encountered two parts to the problem. The first was related to the modelling of automatic control loops of the axes, in motion and when stationary. The result of a single test could relate to the malfunctioning of several components, without it being directly possible to know which one was effectively faulty (in a regenerative loop, should the command or the resulting action be suspected?). The second difficulty related to the complexity of tests to be performed for localizing and identifying the cause of a failure. Use of an oscilloscope and analysis of its signal were practically indispensable for reaching a diagnosis, but only a few maintenance fitters were trained to operate this type of equipment and interpret the results; it was therefore difficult to imagine leaving a part of this type of diagnosis to the operators. The recognition of this problem led to much thought about a different design for automatic control of these machines, one with simple tests which could be carried out by the operators.

These two examples highlight a debate which had until then received little attention. It was becoming urgent to identify more precisely the uses to which the expert system was to be put, as well as its real users.

3.4 Variety of uses, variety of knowledge

It was, however, not all that easy to determine *a priori* what the real use of the expert system would be. Several uses could be envisaged, of which three shall now be described. The first is the one which most readily comes to mind, that of providing the operators with assistance in diagnosing the causes of failures and allowing them either to repair these themselves, or to call upon the most suitable type of technician to solve the problems. The objective would thus be to increase the operator's possibilities of intervention on the facility and to shift the boundary between maintenance and production.

3 Reconstruction of automated knowledge: expertise and usage   167

The second idea consisted of making of it a tool in the hands of the specialists from the central maintenance pool, to formalize and memorize, as they went along, the knowledge gained on new facilities when they received these from the supplier. In a sense the expert system served as a dynamic medium for the creation of technical documentation on machines and their functioning. This consisted of regrouping and formalizing useful knowledge which was then consolidated and enhanced with practical experience gained from failures encountered during the first year of use, together with their solutions. In this case the expert system's objective would be twofold: first, to make it easier for the central maintenance pool technician to adapt and to learn the functioning of a new facility, as he would then be both the expert and the user of the system; and second, to favour the transfer of knowledge on the maintenance of the new machine when responsibility was handed over to the workshop maintenance services. In this case experts and users would be different, but in both cases they would be maintenance specialists.

Finally, a third possible use was proposed by the project leader: the expert system could help to transform and homogenize maintenance fitters' practices by promoting a less empirical mode of reasoning than that which they had generally used for repairing machines. Adopting an approach whereby a logical search for failures was carried out, similar to that of the expert system and relatively close to that adopted by the central maintenance pool specialist, would enable them to increase their capacity to deal with more machines, some of which were relatively new and unknown, or to cope more easily with the technological developments in the new generations of machines. The users would remain maintenance fitters but in a different way, for they would be trained to adopt a more rigorous diagnostic approach than in the past.

These three uses had very different goals. In a way they corresponded to the division of maintenance expertise and to it being attributed to different actors. In these conditions it was clear that an expert system could not satisfactorily fulfill all these missions; the knowledge mobilized and the goals aimed for were, in each case, too different.

It was therefore necessary to opt for one of these possibilities since they all called for a specific course of action. Thus, if the expert system was essentially to help operators diagnose failures, certain questions became decisive; in particular, the exact meaning of diagnosis had to be defined. Which areas of maintenance were the operators to have? Was the expert system to help them diagnose all failures or only those for which they were used to repairing and restarting the machine? Could diagnosis follow the same procedure if carried out by a highly skilled technician or by an operator? On what representation and what understanding of the functioning of the machine could the operators rely? Did the gap between specialists' and operators' technical expertise not make it necessary to reconceive diagnosis in radically different terms? Beyond these conceptual problems, it would be necessary to use only those tests that could be performed by the operators. The impossibility of them being able to use certain tests would make it necessary to invent other methods for obtaining information needed for the diagnosis. But definition of the expert system could well encounter more rigid organizational constraints. Operators' intervention on the flexible cell was limited by levels of authorization; for example, at the time of the project they were only authorized to change fuses and to manipulate weak currents. Perhaps the expert system could have led to a revision of these levels of authorization, yet this would not have been without problems since authorization was defined at the level of each trade. We could multiply this type of question or examine unavoidable issues relative to the two other uses that we evoked; in any event, in order to proceed with the development of the expert system it would have been essential to raise these questions and to choose one of the above-mentioned alternatives. The organizational set-up did not allow the project to perform this change nor to involve future users more closely.

4 The project and the development of stakes: lost relevance

The expert system project has been dormant since the creation of the first knowledge bases and their experimentation. An immediate explanation for this lies in the multiple tasks which monopolized the participants of the project and in particular the maintenance manager. During the development of the system, maintenance of the facility was transferred to the workshop maintenance services. The expert system was not however used as a medium for this transfer of expertise since it was considered that too much preparation by the central maintenance pool specialist would have been required.

It is nevertheless possible to find other interpretations for the situation. We shall first look at the positions of the different actors involved. The system had been presented to the operators and they had discussed it with the expert, but they had never had to use it. Without this experience it is difficult to estimate the true contribution that the system might have made or even the changes that it would have had to undergo in order to be adapted to users' needs. The operators are sceptical about the assistance that the system could have provided; they feel that they know the machine and possible failures well enough to be able to find the causes of breakdowns fairly quickly. Some even claim that their diagnosis is faster than that of the central maintenance pool specialist, whose rational approach is contrary to their empirical one.

The workshop supervisor still wants to see the project followed through and regularly asks questions about its progress. Nevertheless, if its development were to be continued, the maintenance service would have to provide the necessary resources, in particular the availability of an expert.

As far as the maintenance service is concerned, various points of view are to be considered. When using the system the expert did find the same logical diagnostic procedure that he used when he worked on the facility – a procedure which he tried to promote amongst the workshop maintenance fitters. Nevertheless he did regret not having had the time to successfully complete the development of the expert system. Despite the user-friendliness of the software, the writing of knowledge bases remained an extremely long and tiresome task.

The maintenance service managers also claimed to be interested in the project of developing an expert system. They wanted to employ an engineer who would develop methods for stop-gap maintenance and take over the project. However, their experience and analysis of diagnostic and maintenance problems led them to believe that difficulties were related to the availability and gathering of information on the functioning of the machine, rather than to the manipulation of this information for diagnostic purposes. That is why they were very interested in diagnostic cards for collecting this information. Their interest in expert systems increased when it became possible to combine an expert system with these cards so as to limit the number of questions put to users. Finally, they undertook to make the flexible cell as reliable as possible and their action in this respect took almost a year (from the completion of the knowledge bases). Whilst this action led to a significant decrease in the number of breakdowns, the problem of diagnosing failures also presented itself in different terms and was less acute in respect of production.

Finally, it is clear that the project did not succeed in mobilizing and involving enough actors to become effectively credible and achieve the relevance which might have ensured its integration and survival in the organization. It met with several specific difficulties. In contrast to the other expert systems studied in this book, experts and users were distinctly separate actors in the organization. The expert had to formalize his knowledge to make it usable by others. But it appeared that this maintenance expertise was largely dependent on the place occupied in the organization by those who were to use it, as well as on their more general knowledge or their familiarity with the use of sophisticated measuring devices. Thus, the problem of maintenance was not posed in the same terms for central maintenance pool specialists and for workshop maintenance fitters, or for production operators. It would have been necessary to opt for one of the possible uses for the expert system at an early stage and to clarify which of these different actors it was meant to help, since each situation called for a different and adapted response. The meeting of experts and users (around a first model of the system) would have made it possible to define relevant contents for this task, by indicating necessary organizational modifications.

The project did, however, bring to the fore an alternative to the development of an expert system – that of making the facility more reliable. The reduction of causes for failures progressively became a more important objective than the diffusion of diagnostic and maintenance expertise on those same failures. Rather than committing itself totally to the expert system project, the maintenance service undertook to make the flexible cell as efficient as possible. Thus it chose to replace the initial project of diffusing critical but singular expertise, with a reduction of operational problems. It was thereby possible to make do with the expertise fairly readily available to all in the workshop. Moreover, the actual process of enhancing the machine's reliability had considerable potential with respect to learning about maintenance of the facility.

Chapter 3
GESPI
Discovery of station traffic planners' expertise

1 Problem and context: activity in a large railway station

Staff at the French railways, the SNCF, have been using an expert system since the beginning of 1989 to help them control arrivals and departures in a large station. The system, called GESPI (*Gestion Prévisionnelle des Itinéraires*), is the product of an ambitious project which took nearly two years to complete. Before describing the tool itself and the course of the project, this first section provides an understanding of the origin and initial objectives of the project by discussing the problems involved in railway traffic control. It then describes the management of the movement of trains in a large station and the experts responsible for this task.

1.1 The station: a railway traffic node

In a small intersecting station that generally handles very little traffic, the movement of trains and the use of platform lines pose hardly any problems. When a train pulls into the station it is easy to direct it towards the available platform where it will stop for a few minutes for passengers to board and alight, before continuing its journey to the next destination.

The circumstances are completely different when it comes to certain large railway complexes. The station described below illustrates this type of situation in which several difficulties are combined. First, it is a dead-end station, which means that the trains arrive and depart on the same track and must therefore pass one another. But it is also a terminus, which means that tracks are occupied for fairly long periods (in general between ten and twenty minutes for commuter trains and between twenty and sixty minutes for main line trains), the time taken to make up new convoys and for the passengers to board the train. Furthermore, main line and commuter line traffic, with very different characteristics, must be controlled simultaneously. Finally, the tracks are close to saturation and any incident could have a snowball effect, leading to considerable delays. The station's location in the heart of an urban area makes it difficult to extend its infrastructures, so that an increase in traffic must be handled without upgrading facilities. Understandably, in these conditions it is not always easy to find an available platform, or to control arrivals and departures while adhering to both the schedule and safety requirements imposed by stringent traffic rules.

By 1988 the situation at the station we studied was becoming more and more tense and staff were experiencing increasing difficulties in planning and coordinating the movement of trains within the station. In particular, plans for assigning trains to station tracks had become vulnerable, as the slightest delay could disturb the entire programme and necessitate the preparation of an emergency plan.

1.2 A problem which calls for formalization

Solving this station's traffic problems is never simple. The first difficulty is the number of trains running, with over a thousand movements every day. The second is the huge number of possibilities offered by the station; six tracks serve thirty platform tracks, and by means of a large shunting network it is possible to reach almost any platform from any track. In all, trains can have close on 1 400 possible routes. A highly complex combinatorial problem is thus the first characteristic of this planning, the second being the multiplicity of constraints that have to be reconciled. Safety regulations apply mainly to peak hours when there is a train arriving or departing almost every minute. The traffic planners then have to ensure that different trains' routes do not coincide too closely, so as to avoid risks of collision but also to ensure that a train does not have to wait for the track to be cleared (something which could entail a series of delays). Constraints in planning can also be of a technical or commercial nature. For example, directing regular trains and commuter trains to the same platforms every day means that passengers can arrive just in time and expect to find their train on the usual platform. Similarly, to avoid passengers jostling one another on the platform during peak hours, commuter trains should not be scheduled to leave just after the arrival of another train on the adjacent track of the same platform. In practice it is often impossible to accommodate all these constraints -which relate to diverse concerns – and the best possible compromise is then sought.

The preceding description shows that it would be fairly easy to translate this scheduling activity into a problem of optimization subject to constraints, and to develop a formalized model on which a decision making tool might be based. Several attempts have been made in this direction but no research has resulted

2 Expertise in action: the station traffic planners

in a satisfactory algorithm. The major difficulty is a combinatorial one, which has to be solved if the compatibility and consequences of choices of dispatchment or routes are to be evaluated. Because of obstacles encountered by approaches taken directly from operations research, the idea of using artificial intelligence techniques soon appeared as an attractive alternative.

A service at the SNCF Planning and Research Division specifically responsible for promoting the appliction of new technologies, was interested in the potential of artificial intelligence and decided to evaluate it by means of pilot projects. It explored the different possibilities for application, and finally opted for the problem of traffic control in stations, which seemed to be well suited to such a project. The problem was complex enough to warrant the development of a fairly sophisticated tool, and for users to realize that it was designed to help them. It moreover made it possible to implement several AI techniques such as heuristics to by-pass combinatorics, the modelling of constraints and the development of compromises. Furthermore, even though the expertise required had many dimensions (knowledge of infrastructures, traffic rules, schedules, the composition of trains, etc.), it was still easy to identify the experts in the relevant domains and to include them in the development of the system.

From the outset, operations managers and particularly those responsible for running the station were convinced of the advantages offered by the project. Worsening problems of saturation in this station were making traffic control extremely complicated and the prospect of providing staff with a useful tool was in keeping with the policy to improve decision making, particularly with respect to incidents. The project moreover fitted very well into the framework of the modernization of methods and tools, undertaken by the operations managers.

Contact had already been made with a computer service company specializing in artificial intelligence (GSI-Tecsi). It proposed an extremely powerful development tool including all the main AI techniques available (Knowledge Craft) as well as expertise in the development of applications based on these techniques. A mixed team was formed, bringing together engineers from this service company and a computer engineer from the SNCF Research Division, who was to take advantage of this experience to familiarize himself thoroughly with AI techniques. An expert with sound experience in railway traffic problems was to co-operate with the work group whenever necessary.

2 Expertise in action: the station traffic planners

Now that we have seen how and why the decision was taken to embark on the development of an expert system, we need to examine more closely how the SNCF had until then solved its traffic problems. In particular, what was the organizational set-up and what were the skills and abilities of the staff responsibile for this task?

2.1 Three levels of traffic control

To deal with the complexity of problems in assigning platform tracks and scheduling the movement of trains in the station, the SNCF distinguishes three management levels associated with different time-scales:

- *The long-term, or the definition of a service*: On a regional level, the scheduling department draws up a timetable twice a year for passenger trains – the summer and winter services. This document gives the train times for each line or destination but also determines the sequence of trains to be received in a station every day, with their arrival and departure times.
- *The middle-term, or provisional assignment of platform tracks*: For each type of day, the station's traffic planning department studies the assignment of trains to platform tracks, ensuring that there is a possible route for each train between the tracks and the destination platforms. It then draws up a plan for the occupation of platform tracks and a train schedule. The first document indicates the trains which are to use each platform, and the duration, by means of a graphic representation similar to a Gantt diagram. The schedule gives a chronological list of trains which will enter and leave the station, with a track and platform assigned to each one.

There are, however, numerous events which modify these types of days, notably the programming of extra trains, construction or maintenance work which limits the use of certain tracks or platforms, or the modification of tracks upstream. Every day the planning department prepares the next day's documents which describe as precisely as possible all planned activity, integrating known incidents and modifications relative to the theoretical service.

- *Real time, or traffic control*: These plans are sent to the signal cabin to be used by the dispatchers in determining an entry and exit route for each train. When directing traffic the dispatchers have to take care to avoid conflicting routes which force one of the trains to stop until the track is clear and which are therefore a source of delays. Numerous incidents have to be dealt with – especially delays -which prevent them from following instructions on the plan. In such cases a new satisfactory solution, which may differ considerably from that of the planning department, has to be found. The dispatchers have to be able to react very fast -which implies a thorough knowledge of the station, its facilities and expected traffic – and to make an on-the-spot evaluation of the consequences of their decisions. Generally they try to apply the planning department's solution, for at least they know that it is globally coherent. Moreover, the station has to be advised of any change they may make to the original plan, even if this is only for the purpose of displaying it on the arrivals and departures board.

2.2 Limiting initial ambitions

These different procedures are aimed at reducing the complexity of the problem by distinguishing stages to prepare possible options – when this is feasible – and limit emergency decisions to unexpected problems only. But the work and the conditions in which it is performed differ considerably at each of these stages. There is little in common, for example, between the quasi-automatic decision making of a dispatcher to sort out a traffic conflict, and the detailed planning of a service schedule for several months. The expert system could not possibly cover such diverse activities simultaneously; it was therefore necessary to specify which aspect of the problem would be modelled. At first the choice fell on the development of a tool for assisting the dispatchers at the signal cabin level, without excluding the prospect of completely automating the routing of trains if the system performed satisfactorily. However, the essential requirement that the system operate in real time in response to actual events, added a technical difficulty. After several preliminary tests the project leaders decided to avoid this complication by choosing to build a simpler tool for assisting planning management, one which would help staff in the planning department in their task of preparing the dispatchers' work.

2.3 The unseen staff in the planning department

Whereas dispatchers occupy a conspicuous position in the station, the traffic planning department staff seem more modest and unobtrusive and their work less visible and less known. In their offices they calmly prepare the battle plans for the following day, although this calm is often an illusion. Unexpected events occur frequently (e.g. strikes or accidents), forcing them to hastily draw up a new plan for the same afternoon or the next day. Their task is clearly not an easy one. In the department there are experienced people who have been working at the station for ten or twenty years and whose experience was gained in the signal cabin or in operations. According to the supervisor, over a year is required to train a planner familiar with all the subtleties of the job. Some of them know the station and the traffic so well that they are able to identify, almost at a glance, the flaws in a plan. (Even though this document is filled with seemingly obscure information, it is a graphic representation of a timetable for each of the thirty platform tracks, detailed down to the last minute.) The station traffic planners will be discussed in more detail later in this chapter.

3 Reconstruction of automated knowledge: progressive discovery of expertise

The history of the project is too long and rich in new developments for us to follow its progress in detail here. We shall therefore first give a brief description of the technical functioning of the expert system, with an indication of the type of knowledge it uses, before looking more closely at the most significant moments in the life of the project.

3.1 GESPI: knowledge and reasoning

The station's traffic problem can only be dealt with if its combinatorial dimension is limited as far as possible. To achieve that objective the expert system avoids rebuilding a complete plan each time; it tries, like a human expert, rather to use a basic solution developed for the regular part of scheduled traffic. Thus, it starts by checking – taking into account the state of facilities and particularly work in progress – if there is always a route allowing regular trains to reach their allocated platform track, without this leading to a conflictual situation. Next, it tries to assign a platform track to extra trains and to regular trains that have to be relocated (about 250 trains per day) and ensures that there is a route for them to use without creating traffic problems. It then asks the user to manually assign the remaining trains for which it has not managed to find a solution. Finally, it prints out the results in the form of an "occupation of platform tracks diagram" and a "train sequence schedule", identical to those drawn up by hand. Let us now look in more detail at its reasoning.

3.1.1 Finding a route

A route is made up of short sections of tracks, called blocks, between two signals. Since some of these blocks can be assigned to the routes of several trains, traffic rules guarantee safety by precluding the simultaneous occupation of a single seg-

3   Reconstruction of automated knowledge: progressive discovery of expertise   177

ment. One of the two conflicting trains will be stopped and the two corresponding routes will be considered incompatible. It will then be necessary to find, for each train, an alternative route which will not be occupied simultaneously by another train. Should this occur, the dispatchers will be forced to make one of the conflicting trains wait until the other one has cleared the track. The resulting delay may have a snowball effect and disrupt traffic by multiplying unforeseen conflicts. It is not, however, possible to systematically compare all routes to check their compatibility; such a test would simply multiply the combinatorics.

Fortunately not all trains are likely to encounter this type of problem. It is even possible to detect, beforehand, sets of trains which are in a situation of potential conflict. For example, we know that if two trains travelling in the opposite direction have similar times, and if the order of the tracks and their platform tracks is inverted, the two trains will have to "cross" each other. This crossing will not necessarily result in a traffic conflict but it does introduce a risk which should be examined during the choice of their respective routes. About ten rules of this type have been formalized, thus making it possible to identify all possible situations and to define for each train the list of all other trains with which it is likely to be in conflict.

The combinatorics can be reduced considerably by means of these relations of potential conflict which make it possible to break it down into independent subproblems. Thus, if there is no relation of potential conflict between two sets of trains, routes can be given independently to the trains of each of these two sets. Within a set of trains characterized by relations of potential conflict (such a set generally contains about a dozen trains), the one which has the most relations with the other trains in the set is examined first. An attempt is thus made to first assign a route to the most difficult situations, followed by the ones which pose less of a problem. At each stage a check is carried out to ensure that the new route assigned does not conflict with any other route. The system relies on resolution heuristics which, associated with each type of conflict, describe the characteristics of routes which make it possible to avoid conflict.

Thus in our preceding example of the crossing of two trains, it might be safer – depending on the time difference between the two – if their routes crossed away from the station so as to ensure that the common part of the track is freed by the first train when the second one arrives. Supposing that the arrival time of train A coincided with that of train B and that their routes crossed, to avoid any risk of traffic conflict their routes should cross far from the platforms. For if a train takes two minutes to pull into or out of a station, train A would have crossed the signals that it shares with train B almost two minutes before the latter's departure. When train B arrives at the signals the tracks are then clear. If, on the other hand, train A arrived two minutes after the departure of train B, their routes would cross near the platform tracks.

The expert system does not, however, always manage to find a route that is free of potential conflict. In this case two possible solutions exist: slackening the traffic rules, if the conflict is a minor one, or reconsidering a preceding choice. If it opts for the second solution, it will, for example, try to assign a different route to the second train involved in the conflict. This technique is called backtracking. When the system backtracks it returns to the state in which it was at the time of the initial choice (the states relative to all the key steps in its reasoning are memorized). It nevertheless saves data of later options, in particular tested incompatible situations. To avoid excessive backtracking this is limited to the small set of trains considered each time. If, in spite of all its attempts, the system can still not assign a route to a train, it reconsiders the choice of a platform track. The train is then put with a group of other trains without platform tracks, to be dealt with afterwards.

3.1.2 Assigning platform tracks

Once all the regular trains that can possibly be given a route have been dealt with and their platform tracks confirmed, the system has to assign platform tracks and routes to all extra trains and remaining regular trains. In fact the relevent notion for finding a platform track is not a train, but a movement, that is to say, the set formed by a train pulling in and another one pulling out of the station. These two trains are each identified by a number (an even or odd number depending on the case) and a time, but if they are going to use the same carriages it will obviously not be possible to assign two different platforms to two trains of the same movement. The platform track will therefore be occupied by the movement during the entire interval between the arrival of the first train and the departure of the second train in the movement. Occupation time generally varies between twenty and forty minutes, but it can be as much as several hours for a train with sleeping carriages, for example.

To assign a platform track to a movement GESPI first determines all the platforms that are technically accessible from the tracks at the entrance and exit of the station. In general the system has several alternative solutions. Priority rules will guide it in choosing an order for examining possible platform tracks so as to find a solution rapidly, avoid fruitless attempts and so save time. One of these rules states, for example, that if a train arrives at a peak hour and leaves at an off-peak hour, it is preferable to give it a platform that follows on from its tracks when it arrives, so as to limit its movements in the station and thereby all risk of conflict – even if its exit route cuts across several tracks.

The system then checks if one of these platform tracks is free at the right moment and for long enough to receive the relevant movement. If a platform track meets these requirements, the system checks that this assignment conforms to the different technical, commercial and traffic constraints. The following examples illustrate such constraints:

3 Reconstruction of automated knowledge: progressive discovery of expertise   179

- *technical constraints*: the platform has to be long enough to accommodate the entire train; or commuter trains must be assigned to platforms equipped with date-stamp machines;
- *commercial constraints*: the same platforms must preferably always be assigned to the same trains; two commuter trains should not arrive and depart on the same platform at the same time;
- *traffic constraints*: a route must be compatible with the movements of other trains for reaching the platform and leaving it (to check this constraint the system looks for a route, as described in section 3.1.1).

When all these constraints are respected the movement can be assigned to the platform track, otherwise GESPI tries to find a different solution. If the systematic search for a platform track is unsuccessful the expert system ignores the constraint of availability of the platform track during the entire period in which the 'movement' is stationary. It then diverts certain trains, initially assigned to that platform, to other platform tracks. A platform is thus freed for long enough to assign the problem train to it. In this case the expert system starts checking whether it can allocate the freed platform to the movement under consideration, without failing to comply with the constraints. It then tries to reassign the diverted trains to other platforms, preferably as close as possible, along the same principle. It is, however, forbidden from diverting other trains when trying to integrate a train that has already been diverted. The system may well be led to exploring several alternatives for relocating a train, before finding a suitable one. There is another possible reason for diverting a train, which could be used as a last resort. If an assignment is impossible due to an insoluble traffic conflict with another train, and there is no other possible assignment for the first train, the second train which has already been assigned a platform may be relocated to another platform.

It may happen that even by diverting trains a solution cannot be found; the search is then started again but without the commercial constraints. Finally, if the problem can still not be solved, GESPI informs the user who then has to find a solution manually by accepting that certain constraints, for example relative to traffic rules, must be ignored. These situations concern about 1 or 2% of trains.

3.2 The project underway

Table 1 gives the main dates of the project and highlights several stages which overlap rather than following one another chronologically. First the work group, composed of the Research Division's computer engineer and the two engineers from the computer service company, in close co-operation with a specialist in traffic problems in the station, attempted to formalize the problem and model it. In the second phase they developed a tool which applied these principles to obtain as-

Table 1 – The Gespi project: main dates and events

Dates	Stages	Events
Early 1985	Emergence of the project; initial request by station	Research division becomes involved in project
June 1985	Feasibility study; breadboarding (Research Department & GSI-TECSI)	Redefinition of project: from signalroom to planning office. Choice of method solution: – heuristic → algorithm – modify → building a plan
Early 1986	1st development phase – modelling infrastructure – recursive algorithm for searching for routes	
Mid-1986		Start of involvement of future users. Change of expert
Early 1987	2nd development phase – choice of platform tracks – graphic interface Test on 1st prototype: Obstacle calculation time 9 hours → change of material	
August 1987		Decision to industrialize financial package and planning of project. Plan for new organization of station planning department
Early 1988	Development of office operating environment. Fine-tuning edition of station documents	
July 1988		Installation in station planning department
	1st validation by experts on test checks	
		Fine-tuning resolution rules and strategies
	Training users and continuation of validation tests	
		Users contest Gespi: "too many challenges". Logic stability → respecting constraints → rules for determining acceptable conflicts
Early 1989	Implementation of Gespi and the new organization. Daily use first with manual plans	
June 1989	Preparation of service change with Gespi	

signment schedules. Once they had designed the expert system, the core of the application, they had to develop a number of interfaces to allow for its effective integration and implementation. The actual installation of the system in the planning department's offices made it possible to simultaneously run a test phase for validating complete days, and have the users adapt the system to their needs. This fruitful interaction resulted in several modifications in the reasoning and functioning of the tool. During a period of several months the expert system was used concurrently with the existing manual system, before being made effectively operational. The following paragraphs provide a more detailed description of these different phases.

3.2.1 Successive modelling

Having learned a lesson from unsuccessful attempts at algorithmic optimization, the project team tried to follow the experts' line of reasoning as closely as possible without trying to invent a better solution. The approach implied an effort in understanding that reasoning: how did they avoid the traps of combinatorics, how did they manage to break down the problem, what was the sequence of their reasoning, and so forth?

But experts, no matter how good they are, do not generally have a preconceived formalized representation of their line of reasoning and certainly not one that is complete enough to serve as a basis for computer modelling. The latter consists more of an interactive learning process where the initial understanding of the problem and the way it is handled produce a first translation in the form of a model. Tests then make it possible to compare this first model with the expert's knowledge and any discrepancies can lead either to an enhancement of knowledge on the expertise itself or possibly to the principles of the system being challenged. The problem can then be specified or shifted and may give rise to a new development on the basis of a better understanding. Several back and forth steps of this kind are often necessary to progressively build a system whose results are judged satisfactory by the expert. Since it was assumed in the GESPI project that expertise existed and that it would be possible to translate it in a form that was compatible with the computer system, it was necessary to urgently check the validity of these assumptions.

Encouraging results were soon obtained, as the experts expressed their knowledge on reasoning in a form close to the heuristics and rules used in artificial intelligence. We have already seen that reducing the combinatorics meant taking into account potential conflicts between trains. To be able to use this notion, it was necessary to identify such situations as a preliminary step, even before assigning the trains to platform tracks. Small numbers of simple rules provided by the expert made this possible. Several other heuristics were brought to light during these first discussions, notably those concerning the order in which platform tracks should be examined, so as to limit the number of attempts at assigning them and thereby to reduce calculation time.

It was similarly possible to start expressing, in the form of rules, diverse constraints which had to be respected when assigning platforms. The ease with which these could be clarified depended on their technical or commercial nature. Technical rules constituted the basic knowledge; they were largely shared, better formalized and better established in practice than commercial rules. The latter seemed more diffuse; they involved different actors, particularly the sales department, and were less formalized – which explains why they were only expressed and taken into account in a second phase, and then only progressively.

However, a considerable part of the knowledge concerned a description of the objects involved in planning and the relations between them, rather than the reasoning itself. The concepts of trains, routes, tracks, etcetera, presented no problem of definition, but a practical way had to be found to represent and manipulate them in GESPI. AI's form of representing knowledge proved to be most effective for integrating the wealth and diversity of these elements. Thus, a train was described in the form of an "object" by its list of properties (length, type, arrival and departure times, etc.). It was situated in a hierarchically organized arborescent structure, which enabled it to "inherit" from all the properties of the class of trains to which it belonged (main line train, commuter train, etc.). Relations between trains were expressed easily, for example relations of potential conflict were expressed by means of a semantic network between trains.

Another significant task carried out simultaneously consisted of describing infrastructures. The list of 1 400 routes and their composition of blocks was drawn up from station plans. Some blocks were grouped together since this detail was considered unnecessary by the expert, even if the existence of short segments was useful in the movement of trains. All incompatibility between blocks was also identified. This information was represented in the system in the same form as that outlined for trains.

Without describing in detail each of the successive models, it is possible to identify a progression in the way in which the problem was approached and the type of response it called for. The initial idea of a tool which would develop the best possible plan from scratch every time, was progressively transformed into that of a tool which tried to adapt and complete a basic solution. There was thus a shift from a logic of planning to one of replanning. This evolution took place in stages, which led to the system under development being dropped several times and new work being started on a different basis. Owing to the flexibility of AI techniques little time was lost in this iteration. In particular, the principles of representing objects did not vary and reasoning mechanisms, separate from these data, could be written fairly briefly. Each of these modifications corresponded to the intention to stay as close to the expert's line of reasoning as possible. (In fact their mode of reasoning was discovered and written progressively by means of a comparison with

3 Reconstruction of automated knowledge: progressive discovery of expertise   183

GESPI's first results on a test deck.) But another significant event also explains these changes of orientation.

3.2.2 Change of expert, change of outlook

With the system's first encouraging results, its managers thought it would be useful to start involving the head of the station's traffic planning department who was the future user of the system. He was therefore included in the design group, but disagreement between the two experts rapidly became apparent as they had a very different view of the problem and the way in which it had to be handled. These differences can easily be explained if one considers the position of the two experts in relation to the problem of controlling traffic. The first had sound but outdated experience since he had been directly involved in traffic control in the station about ten years earlier. In contrast, the other expert was directly responsible for the daily preparation of diagrams assigning platform tracks and was thus in constant contact with difficulties in preparing and implementing these documents. It was naturally important for him to find in GESPI certain processing principles, notably that of aiming for a certain degree of stability during planning. Another explanation for the experts' differences lay in the changes in the station and its traffic over the preceding ten years. The worsening traffic situation due to saturation had led to a different way of considering the problem. For example, to avoid disturbing a fragile equilibrium, the traffic planners preferred lowering the standards for assigning platform tracks, rather than risking a series of repercussions. The evaluation of what a good assignment plan was, had changed. Other changes concerned modifications to the infrastructure (e.g. equipping those platforms which received commuter trains with date-stamp machines, and thus starting to specialize resources according to the nature of traffic), the profound reorganization of timetables, the appearance and increasing importance of commercial rules, and so forth. The arrival of the new expert contributed to the introduction of profound changes in the system. Certain principles chosen at the beginning were however retained even though they did not correspond entirely to the planning manager's view of the problem.

The question of choosing a suitable expert raised that of why the project leaders had not involved future users from the start. The main reason was the initial uncertainty as to the outcome of the project and the issue of staff reductions that was associated with it. The combination of these two elements led the project leaders to adopt a careful strategy and to opt for a preliminary phase to explore the technical feasibility of the project before involving future users and informing them about all the stakes. This careful strategy nevertheless had the drawback of subjecting the project to a twofold risk: that of increasing the development time and that of creating a discrepancy between the tool and its users' real needs and constraints.

3.2.3 Peripheral developments

At this stage the part of GESPI that prepared plans assigning trains to platform tracks was theoretically useable. However, in order for it to be truly operational a number of interfaces had to be developed to supply the core of the application with the basic data it needed to present its results in an appropriate form. The importance of interfaces in computer applications, and particularly in those using AI, has often been emphasized. Their quality largely conditions the acceptability of the system in the users' eyes, as well as the integration of the tool, but their development is generally an enormous task and consumes a large part of the project's resources. GESPI was no exception and the time taken to develop the interfaces was almost as great as that devoted to the development of the planning tool as such.

The architecture chosen meant that the interfaces were developed in a second phase. The system was installed at a work station, in the heart of the application, but in order to function it required the addition of a whole set of peripheral applications relying on a network of micro-computers to supply GESPI with information (extra or modified trains, lists of work on infrastructures, etc.) and present its results in an acceptable form. Figure 1 is a simplified diagram of this architecture. Graphical representation played an important role, for GESPI showed the assignment of platform tracks or train routes for a given time-frame, so highlighting potential conflict. Its results were meant to have the same form as those prepared formerly by hand, but the planners' diagrams contained a large amount of information which was not directly manipulated by the expert system for drawing up its plan (for example, the movement of engines without wagons). In fact the planning department is situated in the heart of the station's information system and its mission is not limited to the production of assignment diagrams. Additional developments on GESPI made it possible to add this information and to reconstitute the original detail of the documents.

3.2.4 First encounters with users

The question of evaluating the quality and relevance of the plans proposed by GESPI arose very early on in the project. Tests on increasingly complete and representative test decks allowed for a first assessment which was confirmed with successive adaptations of the model. Experts and future users agreed that the results were satisfactory and that they differed very little from those which would have been prepared by hand. They considered, however, that the line of reasoning implemented by the system to obtain these results did not correspond precisely to their own. The recognition of configurations encountered formerly normally guides experts in the identification of tricky situations – on which they focus their attention – and towards solutions for these difficulties.

3 Reconstruction of automated knowledge: progressive discovery of expertise

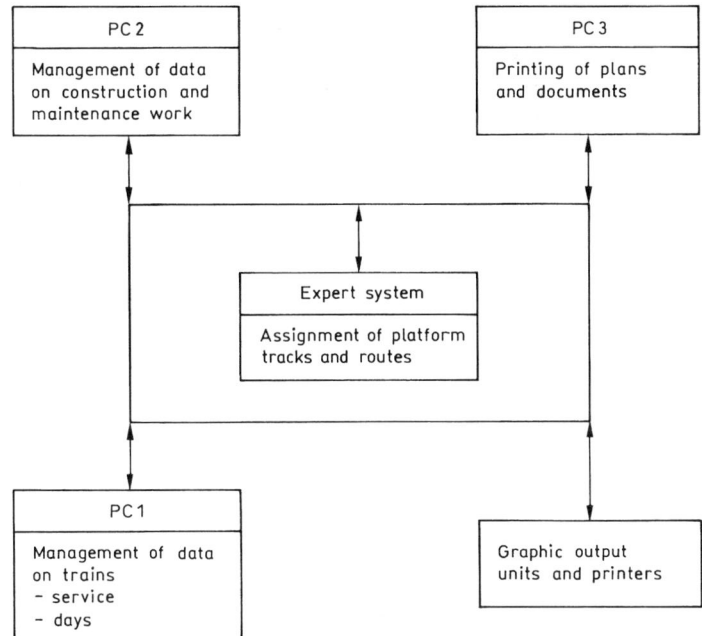

Figure 1 – Architecture of GESPI and office environment

Before continuing the validation of GESPI and particularly before handing it over to its future users, it was necessary to complete the system with the interfaces mentioned above, to make it more manageable. GESPI was, however, installed in the planning department before the end of the development phase. Its users started adapting it by validating test days and they prepared two sets of diagrams daily – manually and by means of the expert system – before eventually handing the work over to GESPI and abandoning the manual work.

At first the absence of explanations by the expert system on its reasoning appeared as an obstacle to the adaptation of the tool, since its functioning remained obscure. For the users to trust the system, particularly during traffic conflicts, it was necessary first to explain its reasoning by reconstituting the path leading it to certain results, and secondly to prepare the plan by hand so as to check that the same conclusions were obtained as those generated by the machine. Comparison of the system's results with diagrams prepared by hand made it possible to fine-tune GESPI's control parameters in order to reduce to a minimum any discrepancies between the two documents. For example, the interval during which GESPI considered that two trains could be in a situation of potential conflict, was reduced from five to three minutes.

The actual design phase of the computer tool was now complete and it was time to hand GESPI over to its future users. The entire computer system was installed in the planning department where the staff started using it to prepare days. At first they maintained the manual system which was then progressively abandoned. The project itself was, however, incomplete and as we shall see in the following section, unexpected discoveries were still to be made on the planners' real work. Some of the problems brought to light in this way led to further adjustments to the expert system. After its first months of operational use an early evaluation of the project can now be made. The main stakes involved can also be identified *a posteriori*, as these only became apparent during the course of the project.

4 The project and development of stakes: rethinking the station?

4.1 GESPI in the hands of the planners

At this stage it is important to remember the project managers' initial ideas on GESPI's use. Incomplete from the start, the system was not designed to replace the staff in the planning department by automating all their tasks. It was meant rather to help them by relieving them of certain physical tasks such as drawing diagrams, and by proposing the assignment of platform tracks to trains. They were to remain responsible for all adjustments and for dealing with the most serious traffic conflicts. Above all, the planning department remained entirely responsible for all documents drawn up by it.

Furthermore, the problem of the movement of trains was situated in a broader context than that of the station in the strict sense of the word. The optimization of station traffic required a mode of reasoning which went much further, to include constraints and events far upstream. GESPI had, however, only been designed to work on station facilities as such, which made it necessary to complete its expertise with more global knowledge of the railway system.

4.1.1 Processing a day

These principles gradually led to a different organization of work in the planning department, which we shall now describe briefly by outlining the progress and sequence of steps in the preparation, processing and printing of documents for a day's traffic.

– *Preparation concerns:*

1) extracting from the database the train service for the given day, by means of selection criteria (e.g. type of day);
2) constituting movements;
3) updating the list of trains with information on: cancelled trains; changes to the timetable, the tracks, the constitution of movements (number and type of trains); integration of extra trains;
4) inputting data concerning current or planned work on the tracks and concerning the periods that the corresponding facilities will be out of use. Several members of the planning department share this task which can take up to an hour and a half.

– *Using GESPI to process the plans for a day, includes:*

1) loading the files from micro-computers onto the workstation;
2) dividing the trains into two sub-sets: those which are assigned *a priori* to a platform track and those which are without one;
3) updating the list of possible routes, taking into account information on maintenance work;
4) checking whether trains in the first sub-set have an access route;
5) finding a platform track with a compatible route for trains in the second sub-set;
6) displaying on the screen the list of trains which have not been assigned to a platform and for which GESPI did not manage to solve traffic conflicts without disregarding technical traffic constraints;
7) the processing of "rejects" by the planners who forcibly allocate platforms manually. As an aid they use the visual display of the relevant trains' routes.

From the moment the work session is started it takes about twenty minutes to load the files and between thirty and sixty minutes of processing to obtain a first proposition. The interactive processing of rejects can be relatively brief – around fifteen minutes if there are only a few.

– *The printing of documents:*
   After the plan has been validated, it takes another twenty minutes for it to be printed on a graphic output unit. The results are archived at the same time.

In total the processing of a day's plans takes at least two hours work.

4.1.2   New discoveries on the traffic planners' work

We have already shown how the initial conception of the problem evolved by means of a comparison between the first models and the experts' experience. It was, however, only when GESPI was actually put into use that other characteristics of the planners' work became evident. Thus, a discrepancy emerged between former practices and those implicit in the software, even if they had not been defined. Certain adjustments to GESPI made it possible to reduce these discrepancies, but others were not considered feasible at that stage for they would have called into question the very principles of the system. We shall now examine four relevant aspects of this point of view.

4.1.2.1 Incremental adjustment or grouped replanning?

Very early on in the project, the idea of building a complete plan each time was replaced by that of replanning and adjusting a basic solution. This strategy nevertheless left the way open to different ways of using the system. With GESPI, adjustments for a single day were grouped together and put aside until the time came to draw up a final plan for the day. They were then all taken into account simultaneously. In contrast, the planners were used to starting preparations for each day long before the time. They had large plans on which platform tracks had already been assigned according to available information. The five working days of the week were grouped together on the same plan, so that similarities and differences between these days were apparent. As new information arrived in the planning department (e.g. extra or cancelled trains, change of timetable or composition of trains, expected maintenance work), the staff added it to their plans and tried to adapt their assignment of platforms accordingly. A large number of these changes were of course only known at the last minute and could not be taken into account until the day before, but this practice did nevertheless allow the planners to learn a great deal about possible situations and their solutions, which proved extremely useful in dealing with last-minute changes and solving potential traffic conflicts.

There were thus two different types of reasoning. One was based on incremental knowledge and on local and partial solutions, the other implied complete knowledge and the consideration of all modifications. The former, besides the educational function mentioned above, enabled planners to negotiate directly with their interlocutors on certain adjustments, depending on their expected effect on the assignment of platforms, visible at a glance on the plan. The latter left a larger combination of solutions open. By allowing for choices to be made simultaneously rather than progressively, it created the possibility of producing a better plan, or at least of improving coherence. It did not, however, make it possible to answer questions from other services on the assignment of platforms for a particular day in the future.

4.1.2.2 The service: a fairly complex concept

The possibility of starting with a basic solution which GESPI would adapt by integrating the modifications of a given day was based largely on the notion of a service and of typical days. Once the planning department staff had entered all data on trains in the summer and winter services, it was easy to obtain typical days, by

4 The project and development of stakes: rethinking the station? 189

selecting trains on the basis of certain criteria. Nevertheless, on closer examination the notion of a service appears more complex. The change from a summer to a winter service is accompanied by transitional phases which in themselves imply changes, or at least adjustments. For example, on main lines the service change takes place around the 10th July, whereas suburban trains run every half hour instead of every fifteen minutes from the 1st July. Business trains do not run in July and August, but they have to be scheduled at the beginning and the end of the summer service. Furthermore, the conditional criteria for running certain trains tend to multiply. This complexity explains why, at the time of the winter/summer changeover in 1989, the planning department had to resort to manual planning for a while. The need to prepare these transition periods carefully became apparent; this consisted mainly of anticipating and gathering information earlier, so as to enter it into the service's train database.

4.1.2.3 Processing rejects: acceptable conflicts

At first GESPI was unable to assign between twenty and thirty trains out of 950 per day. This number was considered too high by the users but the system's credibility improved considerably when it was reduced to four or five. The improved result was obtained by accepting less stringent constraints and the only ones that the system may still not transgress concern traffic rules. In certain situations, the diagnosis of potential conflict which prevents GESPI from assigning platforms can easily be rejected by the planners. This is due to the absence in GESPI of a model for shifting trains in the station. For example, when a suburban train risks conflicting with a commuter train, the former, being faster, can clear the tracks or platform quickly, without any effective conflict. Similarly, if an engine without carriages arrives in the station, it will be planned like a movement, but will have to give right of way to a conflicting train because it does not have the same time constraints. Certain conflicts are overlooked from the start. That is particularly the case where the proximity between two tracks makes it impossible to use them simultaneously. On the other hand, in certain cases where there is no route the planning department staff can request permission to use a service track for access to a platform. There are some conflicts which cannot be solved; the planners then agree to downgrade the traffic locally by choosing an assignment which might require a train to stop. The expected conflict is not, however, a certainty since everything will depend on the real situation; if one of the trains is slightly early or late the conflict may not occur.

4.1.2.4 A service and a day: two different logics

GESPI is designed to construct an initial model assigning all the trains of the service to platform tracks, or to adapt this basic model to modifications encountered daily. In both cases it uses the same knowledge, for it was implicitly supposed that the knowledge used in replanning was the same as that used in planning. However, the planning department's attitude to these two types of problem seems to be somewhat different. For a given day, their objective will consist of modifying the basic model as little as possible. In certain cases, they prefer downgrading the quality of their planning by leaving a train on the initially planned platform, despite a risk of conflict, rather than assigning it to a platform in a different area of the station.

Although this choice may at first seem surprising, it is perfectly logical considering the consequences that a change of platform would have. Passengers used to taking their train at a specific place in the station would risk missing it if it were on a distant platform, besides which other services, such as the one which displays schedules, would have to be informed. Moreover, the planning department generally tries to find a compromise and accepts the drawbacks, in terms of ignoring constraints, of a strategy that aims at interfering with the basic plan as little as possible.

With GESPI the initial idea of optimization has finally given way to an approach in which experts' practices are copied more closely. Its strategy now consists of relocating a diverted train to the platform closest to the one originally planned, so as to limit disturbances and changes with respect to passengers' habits. This strategy and its reasons were overlooked at first. To obtain the desired result it was simply necessary to limit the number of levels on which an assignment could be challenged and to accept that any constraints, bar those concerning traffic rules, could be ignored. This is, however, a roundabout method and the expert system is not truly capable of coping with a compromise between the different constraints, which would give the user the choice of favouring one aspect or another.

When it comes to the service itself the planners' aim is very different. They try above all to find a plan which will withstand all possible mishaps, and therefore try to leave as much leeway as possible. These two situations obviously require different strategies. The introduction of specific rules or the modification of control variables should make it possible for GESPI to cope better with both situations. This is in any case one of the possible ways of improving the expert system.

4.1.3 Rigidity in interfacing

Despite the significant effort devoted to problems of interfacing, manipulation in real situations has highlighted several difficulties. The first concerns the interactive processing of rejects. Since GESPI does not specify its reasons for rejecting a train, users wanting to assign a platform manually have to reconstruct the system's line of reasoning and from there to imagine a solution. When they have relocated the problem train, they will again be able to rely on the system to insert or shift a train in the plans. Yet GESPI's verification of the new assignment's compatibility with the constraints modelled in the knowledge base may well prove to be problematical. GESPI refuses the assignment if one of the constraints is not respected. It is however precisely in this type of situation that it is impossible to respect all constraints and planners would have to choose a particular assignment advisedly; even though their solution may violate a constraint, it is better than all the others.

The second difficulty relates to the roundabout way in which planners have to introduce an extra train. It is presently not possible to introduce a new train directly on the workstation being processed. The information first has to be entered into the micro-computer, the modified file loaded onto the workstation and the session restarted, all of which takes some time. Research is underway to streamline this procedure.

Finally, it is not possible with GESPI to isolate a situation or a local problem in order to analyze it in detail or give a quick answer to a question on the effects of an adjustment. This incapacity to reason locally makes processing clumsy and limits possibilities of simulation. In particular, it may turn out to be a hindrance if the planners are asked by another service in the station to integrate a modification. In such cases the interlocutor expects a quick reply, which the planners cannot give without reprocessing the entire day.

4.2 Evolution of the traffic planners' job

One of GESPI's most obvious impacts lies in the transformation of the traffic planners' job which its use necessitated. We have already seen to what extent work in the planning department requires varied skills: knowledge of the infrastructure, of railway techniques and of the different departments and services in the station; experience with traffic and schedules; the ability to learn about situations and problems encountered when planning assignments; the capacity to anticipate the consequences of options taken when drawing up a plan. Some people demonstrate amazing virtuosity in the manipulation of these diagrams; they know the characteristics and schedules of all the trains by heart and can list from memory all the main problems encountered when assigning tracks.

GESPI demands that planners acquire new skills, particularly on computers, to operate this relatively complex system. With its introduction they now master two types of expertise, but the actual activities of the department will also be affected. On the one hand, the balance between tasks will be shifted as the time devoted to data input increases considerably, while the tasks of preparing the diagrams are almost totally eliminated. On the other hand, while planning itself remains a significant part of station traffic planners' work, it has also evolved and now consists essentially of checking and validating plans, and of completing them by sorting out problems of assignment which the system cannot solve.

This evolution leads to the following remark: checking GESPI's results seems to its users to be more difficult than designing a plan themselves. The question then arises of whether or not these two activities mobilize the same expertise? Furthermore, if specific expertise for evaluating the quality of the plan is not identified and mastered, is there not a risk of the planners being forced to reconstruct the plan mentally, either for evaluating the assignment of tracks or for finding tracks for a train that is difficult to place?

It also raises questions concerning the training of future personnel of the planning department. How are they to learn the basics of the trade, since these are now incorporated into the system and since they only need to intervene in the most problematical situations? Perhaps they should again be given the task of preparing the plans manually, even if this is only for training purposes, or perhaps it would be possible to use the system's simulation capacities for training, starting by examining characteristic situations with its aid?

Whatever the case, the station traffic planners' job remains valid in so far as GESPI only automates their work partially; they retain the possibility of modifying plans proposed by GESPI, are responsible for resolving difficult cases of potential conflict, and remain responsible for the quality of the work.

4.3 Multiple facets of project evaluation

Over and above the modifications made by GESPI to the planners' job, the first months of operational use have made it possible to propose an evaluation of the project, of its real impact and apparent prospects for evolution.

4.3.1 Control of AI techniques and running this type of project

The first aspect concerns the technical and in particular the computer dimension of the project. We note here that, above all, the initial objective of using AI techniques to assist in the preparation of diagrams for assigning trains to platform tracks has been entirely achieved. For the research division the project also provided an opportunity to evaluate more precisely the contribution of AI techniques and to master these more effectively. This objective has also been attained, although the result is partial since only one computer engineer from the research division has been trained in the use of these techniques. On the other hand, the GESPI experience has highlighted several critical points in the management of this type of project, notably that of the choice of an expert, of an opportune moment for involving future users, and of the necessity to opt either for a project which aims at proving the feasibility of an approach, or for one which aims at creating an operational tool.

4 The project and development of stakes: rethinking the station?   193

Another lesson learned by the computer engineers of the research division concerns the architecture of the system. It seems, in retrospect, that with greater integration of data management the efficiency and flexibility of the system could have been enhanced. There have even been suggestions to totally invert the perspective and rebuild the system by placing data management at the core of the application and considering the expert system only as a module to be used when needed. The mission and tasks of the system would thus be transformed. They would no longer consist of preparing or modifying an entire plan, but of dealing with specific situations or of checking the coherence of a plan drawn up by the user. The expert system's position at the heart of the application reflects the perception of risks at the start of the projet, when the main unknown factor lay precisely in the feasibility of such a system. It was logical to clear such uncertainty as soon as possible, even if that implied building a relatively isolated instrument to be integrated into its environment at a later stage.

But the aim of the project was not only to provide the research division with an experimental field and to evaluate the relevance and effectiveness of AI techniques. It could moreover never have been developed so far if the station managers had not been convinced of its relatively short-term returns on capital invested. What economic evaluation can be made after a few months of operation?

4.3.2 Returns on investments: reduction of personnel and automation of peripherals

The project involved considerable financial resources. Development costs are estimated to be close on 4 million French Francs, including the purchase of the software Knowledge Craft, the fees of the service company GSI-Tecsi and other service suppliers (e.g. interfaces, peripheral applications), as well as the time spent on it by the research division's computer engineer and by the relevant experts. Over and above this amount, 800,000 Francs were spent on computer hardware.

The first part of the financing was provided by the research division for the purpose of experimenting a new technology, while the equipment was purchased by the transport division. The latter decision was made on a regional level, on the usual basis of profitability. This was in turn calculated on the basis of two redundancies in the planning department, which would allow for a return on capital invested in less than three years.

It is noteworthy that the profitability of the project was achieved not by the added intelligence of the expert system, but by the automation of a certain number of peripheral manual tasks and in particular the improvement of the presentation of the planning department's documents before these were disseminated to the other services in the station. Moreover, as far as the processing and printing of the documents is concerned, the application is not limited to the results produced directly by the expert system. The project provided the opportunity – at the cost of additional specific developments – for automating the production of all documents managed by the planning department: diagrams for assigning trains to platform tracks, train schedules, work sheets, and so forth. In fact the process had been initiated well before the start of the GESPI project and train schedules were processed by computer as early as 1982. This initial database provided a sound source of descriptions for GESPI.

The above economic evaluation is, however, confined to easily quantifiable elements and does not include GESPI's impact on traffic, on the quality of plans proposed by the expert system, or even on the modification of the planning department's work and its relations with the rest of the station. These aspects will now be discussed.

4.3.3 The quality of results: optimum or acceptable solution

As far as the quality of diagrams produced by GESPI is concerned, we have already indicated that it was not particularly "better" nor even different to what had been achieved manually. GESPI did not therefore provide a solution as such to the problem of traffic fluidity in the station; instead it confirmed the earlier diagnosis of saturation of the station, which makes the control of traffic a highly complex matter if the objective of regularity is to be achieved.

The notion of quality of an assignment plan is, however, not simple, for it is essentially dependent on the experts' assessments. Without an indicator making it possible to define and to choose a solution that is "better" than the others, the experts, imitated by GESPI, are content to find an acceptable solution without trying to find an "optimum" one – for which there is no criterion anyway. During the project an attempt was made to find a suitable indicator for evaluating the quality of a plan. One idea consisted of measuring the number of delays in a single day but the drawbacks of this indicator rapidly led to it being abandoned, without other possibilities being considered as promising. It is in fact an *a posteriori* indicator which evaluates the outcome of a plan and which cannot be used to judge its value *a priori*. Furthermore, such an indicator is not easy to interpret, since a delay may be the result of the plan, but it may also be due to an incident upstream from the station.

The question of evaluating the quality of plans implies that the goals and reasoning which guide their development should be identified beforehand. But the progress of the project shows that these were not clear at the beginning, not even in the experts' description of their activity. Even though the idea of optimizing the plans was rapidly abandoned, the expert system designers nevertheless tried to do away with potential conflicts between trains, thereby favouring a logic of respecting constraints, which led them to modify initial plans until they found a conflict-free situation. This strategy was moreover confirmed by the experts' attitude. When they were asked to analyse conflictual situations between trains they

4 The project and development of stakes: rethinking the station?   195

always managed to come up with rules enabling them to avoid this type of conflict, even if in their daily practices the potential conflict – replaced in its context – might have been considered acceptable.

It was only when the users found themselves in the situation of having to implement GESPI's plans that their logic became clearly apparent. They then contested the results obtained by GESPI, claiming that these generated too many adjustments compared to the basic solution proposed by the service. To avoid changing the basic solution too much they accepted that certain constraints could be ignored, even if this introduced a risk of conflict and reduced the "quality" of the assignment. The planning department's role in the station as one of stabilization and absorption of mishaps became clear. It was therefore necessary to modify GESPI to make it more consistent with the planners' logic. To that end rules distinguishing acceptable conflicts and inessential constraints had to be determined.

In the end, from the line managers' point of view the system's contribution is to be sought elsewhere, and not in the improvement of results. It lies rather in the automation of the management and printing of all the service's documents – i.e. in all the peripheral office applications as compared to the expert system as such – on the one hand, and in the greater capacity for negotiation with other services, based on simulation results given by GESPI, on the other hand.

4.4 The planning department at the centre of the station and its changes

One of the effects of the project has been to focus attention, throughout its duration, on the planning department and its staff who were hitherto unobtrusive actors in the station. Their activity and role in the station are now better understood and the difficulties with which they are confronted have become apparent. But beyond highlighting problems, GESPI has opened new vistas for these actors and initiated new thought on the developments it could support.

The project has made it possible to measure the intensity of the problem posed by the saturation of the station. GESPI, in spite of the hours of thought by the engineers who designed it, has not been able to do any better than the planners themselves. It seems quite likely that no better solutions than the current ones exist. On the other hand, there is no certainty that the planning department will be able to continue absorbing, as in the past, the inexorable intensification of traffic without a thorough upgrading of the station's facilities. GESPI may then prove to be very useful, as its simulation capacities make it possible to test planned changes, to point to the most critical areas and to guide the design of new facilities.

The planning department's role is, however, not limited to the preparation of assignment plans; it also manages numerous technical databases which are in a sense the basis of the station's computer system. The department is at the centre of discussions between different services and the implementation of GESPI can therefore help to enhance its relations with other services by providing additional information. Its use can, for example, focus attention on trains which are systematically difficult to place, and provide planners with the information they need to negotiate a change of schedule or of tracks. It can also enable them to determine the quantity of acceptable maintenance work on the tracks. Until now the planning department was not equipped to assess the potential impact of such work on the fluidity of traffic at a given time, and all proposed work was accepted. The repercussions of planned maintenance work can henceforth be simulated on GESPI. Simulation results could lead to proposals to spread such work out over time in order to limit its adverse effects on the flow of traffic in the station. Finally, in the case of a major accident GESPI makes it possible to draw up a new assignment plan in two to three hours and to distribute copies to the different actors in the station, who then have to deal with the new situation.

The databases which GESPI uses could moreover serve numerous applications in the station, some of which are presently being examined, e.g. direct supply of information to timetable display screens, announcements by voice synthesis, and so forth. GESPI will henceforth be oriented to the management of all types of useful information, with the preparation of schedules and of diagrams assigning trains to platform tracks being only part of its functions. The planning department's focal position in the station will thus be enhanced.

4.5 Prospects for diffusing GESPI

Problems of fluidity and regularity of traffic do not exist in this station only. Most large stations are confronted to some degree by the same difficulty. At the start of the project the generality of the problem and prospects for diffusing the expert system in multiple copies were amongst the main elements in the choice of a subject. What would an adaptation of the system to other sites imply today? Which elements are common and which are peculiar to the station used for experimenting GESPI? In other words, what is the system's strength in respect of a change of station?

Clearly, even if the principles of description and reasoning remain valid, it would still be necessary to reconstruct a large part of the system. This would of course concern the description of infrastructures but also, more fundamentally, the exact nature of problems encountered in traffic control and the strategies used to solve them. In certain stations, for example, there are very few platforms and control of the flow of passengers is more important than that of routes and traffic conflicts. The game of rules and weighting to translate these priorities can be very different from one station to another and would require in-depth explanations and formalization with the experts every time. Despite this handicap, several stations have shown their interest in the experiment and would like to have a similar system.

Chapter 4
Naval
Undefinable expertise of strategic planners[1]

In October 1985, the directors of a large multinational oil company decided to launch the Naval project with the aim of developing a computerized strategic planning tool. This tool was to incorporate the general principles of artificial intelligence and, in the context of the group's offshore oil wells, was to make it possible to programme the utilization of complex and costly equipment.

The reasoning behind the Naval project was twofold. It comprised the research logic of the group's AI specialists, who wanted to investigate the technique's potential to build action plans, and the development logic of the operations managers in charge of oilrigs. For the latter, the economic stakes associated with the management of rigs and the tension which had been experienced shortly before, more than warranted the development of an instrument for streamlining their planning. Each of these dimensions, while contributing towards the legitimacy of the project, was nevertheless accompanied by a great deal of uncertainty. On a technical level the "intelligent" generation of programmes still had to be demonstrated, whilst on an organizational level the insertion of the tool in a process of strategic decision making between management and autonomous subsidiaries raised a large number of questions. That is why, from the outset, the project's promoters wanted the technical work of the system's designers to be complemented by a support team of participant observers. This mission was entrusted to the authors who were to participate in the design of the system and to monitor the problems that arose with the insertion of the tool into the life of the company. Whilst this position allowed us to be stakeholders in the technical development of the tool (on the level of operational specifications) it also enabled us to give our constant attention to the project. Throughout its course we were able to question the relevant managers and suggest options or guidelines which they then discussed with us. We could thus be actors in certain developments, or the witnesses of decisions which sometimes had to be interpreted, but the effects of which could be felt in the life of the project.

---

1 A short description of this case has already been published under the name "The MET-AL case", in Michael Masuch (1990). "Naval" is the real name of the project.

An account of the dynamic process to which we contributed will now enable us to consider several aspects of the project and the context in which it took place.

1 Problem and context: oil exploration and instability

1.1 Drilling: decentralization, uncertainties and commercial negotiations

An oil company cannot be content to exploit a set of known oilfields, for it has to constantly think of replenishing its reserves and try to find new and more cost-effective deposits. Exploring and developing new fields is a long, complex and extremely expensive process which may span an entire decade and mobilize hundreds of persons. It also involves a high risk factor and may be interrupted after the initial attempts, should these prove to be discouraging. Drilling occupies a critical place in such ventures since it makes it possible to confirm the presence of oil and so guide decision making.

The events and processes examined here start with the localization by a geologist of a prospect, i.e. an area in which the geological structure indicates the presence of oil, in conditions which make it possible to envisage exploitation. Negotiations are then entered into with the host State in which the prospect is situated, and lead – if all goes well – to the signing of an exploration license authorizing the company to prospect in that area during a given period. The contract also defines the conditions of exploitation in case of a discovery. Concessions are not in fact granted to a single company but to consortiums in which the risks are shared and of which one company is made responsible for technical operations. In spite of the sophistication of geophysical analyses, it is not possible to determine the characteristics of an oilfield in advance; these assumptions must be verified by drilling appraisal wells at specific points of the supposed structure. Drilling constitutes the most expensive part of exploration programmes and, despite the increase of geological and geophysical knowledge, remains a high risk investment. Professionals estimate that under 10% of the wells drilled during an exploration programme lead to a cost-effective discovery. Furthermore, if the drilling is offshore costs are even higher and rigs with specialized teams are required.

These conditions help to explain the complexity of decisions to be made. What is a valuable prospect? How can a drilling programme be justified? When will the results be considered encouraging? Answers to such questions are subject to a multitude of uncertainties. Over the past few years the evolution of crude oil prices due to successive oil crises has certainly been one of the key variables in the economic evaluation of prospects. Yet even without such fluctuations, evalu-

1 Problem and context: oil exploration and instability 201

ation of a prospect is still largely dependent on changes in the group's exploration policies. During the past few years only half of all planned drilling was actually carried out, with the other programmes being replaced by new and more promising prospects that had appeared in the meantime.

Generally, the rigs and drilling teams do not belong to the oil company, but are leased for a given period or for a certain number of wells agreed on by the company together with the contractors specialized in running fleets of rigs. An oil company has the choice of a large variety of rigs suited to the different drilling conditions and characteristics of the prospect. This is in fact an oilrig market, with a balance of supply and demand between the rig contractors' and the oil companies respectively. Prices and conditions for renewing contracts depend on the uncertainties weighing on the exploration programmes and the rate of use of the machines. For any given oilrig and the specialized team operating it, the price varies between 50,000 and 100,000 dollars per day, depending on the period. These markets are concentrated in the main oil-producing regions. Machines can be transferred from one area to another, but such transfers immobilize the machine for several months and are therefore relatively rare. Each of these regions can thus be considered by the companies as relatively independent.

Multinational oil companies like the ones discussed here are mainly active in the countries in which they have subsidiaries. In principle, these subsidiaries enjoy a large degree of autonomy, justified by the fact that they all finance their own exploration. They decide on their exploration programmes and negotiate the rental of the equipment needed to implement them. In practice, however, the subsidiaries' autonomy may well be limited by the extensive negotiation network in which they find themselves. They have to define their exploration programmes in relation to several local actors, i.e. the government of the host State, the other partners in the same venture, and the oilrig contractors. These programmes are moreover subject to the budgetary and technical control of several head office divisions. (e.g. verifcation of the geological value of the prospects, the compatibility of the rigs with the characteristics of the prospects to be drilled, or the legal aspect of the rental contracts). The experts in charge of this control have a fairly complex role: on the one hand they provide assistance to the subsidiaries when requested and, on the other hand they advise the group's directors during negotiations – particularly budgetary – with the subsidiaries. On a lower level, regional divisions co-ordinate the activities of subsidiaries in a single geographic area.

Examining the peculiarities of drilling operations leads to at least one conclusion: the planning of this activity is not a simple matter of temporal co-ordination; it implies the reconciliation of multiple strategies in the face of uncertainties, all in the framework of a network of external negotiations. The nature and means of this co-ordination have evolved over the past few years and this evolution, which we shall briefly retrace, explains the origins of the Naval project.

1.2 The stand-by crisis of 1982

The second oil shock in 1979 took place in a context of existing tension on the offshore oilrig market. The worldwide utilization of these facilities was at about 90%, which meant that there was very little room for immediate expansion of exploration programmes. Tension was therefore extreme in 1981. All facilities were occupied, and this at a time when funds made available by the tripling of the crude oil price were an incentive for companies to expand their drilling programmes. Moreover, the anticipated increase in the price of crude oil intensified the phenomenon by guaranteeing the profitability of increasingly difficult or uncertain prospects, so making them seem more attractive. This classical expansionist circle was abruptly challenged at the end of 1981 with the reversal of the oil price trend and the resulting stagnation of exploration programmes.

During 1982 the crisis set in. New equipment, for which production had received the go-ahead during the enthusiasm of 1980, arrived on the market whilst the rate of utilization had dropped to 80%. The market was saturated. Furthermore, during the period of peak demand rig contractors had managed to negotiate long-term contracts (from 1 to 4 years) for the new equipment, at very high rates. The oil companies therefore found themselves with rigs which they did not need and which continued to cost them large sums of money. The situation varied from one subsidiary to another, but some found themselves in considerable difficulty when the local authorities or their partners refused to share the expenses incurred by unutilized equipment. Moreover, the transfer of machines from one subsidiary to another also became difficult to negotiate since there was an ample supply of machines at low prices on the market.

2 Expertise in action: planning without a planner

2.1 Response to the stand-by crisis: strengthening co-ordination mechanisms

During the expansionist period which followed the oil shock, the group's parent company (with which we worked on the Naval project) played the important role of assisting its subsidiaries in their frantic search for available rigs. Task forces were formed to scan the market and seize the slightest opportunity. On a regional level, they tried to favour the sharing of rigs between several subsidiaries by co-ordinating their activities. After the reversal in market trends when the problem of stand-by (inactivity of rented machines) arose, these teams continued playing a similar role but with new objectives. Their aim was to try to reduce the number of stand-bys as quickly as possible, particularly by avoiding new contracts when idle machines contracted by other subsidiaries could be used. The situation was thus somewhat different during these two periods. During the first, the subsidiaries requested assistance from the central services, whereas in the second period the objectives came from head office, which had to explain and sometimes even impose its co-ordination strategies. The complexity of this situation appeared rapidly, as questions arose concerning the ease of transferring a rig from one subsidiary to another, or the rate that had to be taken into account (the rental contract rate was sometimes twice as high as that on the current market). Highly controversial technical or economic arguments, or those linked to local contexts presented by different actors, led to multiple conflicts between subsidiaries and head office, or between subsidiaries themselves. For example, a subsidiary which hoped to rent a new oilrig at the market price would emphasize the fact that the rig was more efficient and better suited to its drilling programme than those on stand-by, whilst head office would tend to doubt these relative advantages and would recommend that the overall interest of the company be given priority.

Corporate management slowly became aware of these problems and decided, in mid-1982, to strengthen its co-ordination structures. A standing committee was created for rig management and placed under the responsibility of the manager in charge of dealing with rig-related problems. Representatives from both head office and the subsidiaries attended this committee's monthly meetings prepared by a permanent secretary who also kept a watch on the oilrig market. In the week preceding each of these meetings the secretary collected the subsidiaries' exploration plans, as well as new or altered information and the opinions of the central services at head office. He then used these data to update overall plans for the use of rigs, subsidiary by subsidiary. These documents served as a basis for committee discussions. Each rig was examined with respect to its legal situation, the prospects which could be assigned to it and transfers to other subsidiaries. Each subsidiary's programme was also examined and compared to those of other subsidiaries in the same geographical area. The standing committee's objectives were clear, it had to free the group as soon as possible from its long-term contracts, introduce a balance between resources and needs, and minimize the cost of stand-bys. Its creation brought into effect a new system of planning.

Yet this first co-ordinating body was soon criticized by its president. He doubted the reliability of the information transmitted to it and felt that the absence of senior managers from the subsidiaries limited the possibility of effective negotiations during its meetings. Moreover, the risk seemed great that the programmes produced would merely synthesize previous decisions.

Two new measures were therefore taken in 1984 and 1985 to improve co-ordination. One consisted of a rule limiting the subsidiaries' powers to commit themselves in respect of rig rental – any contract of more than six months' duration was to be subject to head office approval. The other aimed at enhancing the standing committee's meetings by a regional body on which the subsidiaries were represented by top level managers. Negotiations would be held every three months on the main trends of their exploration policies.

However, the financial allocation of stand-by costs posed such problems that it had to be dealt with in a specific way, independently of the bodies described above. The principle of total responsibility of each subsidiary for its own decisions was impractical for several reasons: financial and accounting problems, opinions of local partners and, in particular, conflicting decisions during peak periods. A protocol for allocating stand-bys to subsidiaries was therefore prepared by the production counsellor of the relevant geographical area – who seemed to be the right person for finding a compromise between several subsidiaries – and finalized at the end of 1982.

The extensive system developed between 1982 and 1985 to examine and co-ordinate operational means for drilling was based on two objectives: on the one hand that of achieving greater transparency of decisions taken on the different sites, which implied collating essential data; and on the other hand that of setting up bodies for counter-expertise, negotiation, and co-ordination, which recognized the potentially complex and often contradictory nature of each actor's strategies.

In principle this system was an effective answer to the problems peculiar to oil-rigs. Without going so far as to give head office total control over all equipment, it established progressive centralization based on negotiation and compromise. In order to better understand the planning process, it is necessary to briefly describe the techniques which it used and which helped to structure it.

2.2 Planning with very few tools

Clearly, the tools used during the planning process remained fairly limited. Three categories can be distinguished, according to the time scale considered:

- *Exploration plans* recorded, on a group level, the different prospects that were envisaged for all the subsidiaries, with several factors taken into account (e.g. licenses, geological data, partnerships, etc.). These plans articulated a three-year forecast and an annual programme. They were the product of a two-way exchange of data between subsidiaries and head office involving discussions on the prospects of a deposit, and were totally unrelated to any consideration of available rigs.
- *Budget forecasting procedures* translated subsidiaries' annual exploration programmes into budgetary terms. These programmes were composed of the list of oil wells to be drilled for each prospect. The expenses associated with each of these wells were calculated from mean estimates of the price of renting equipment. However, this type of budget was out of touch with reality since it did not for example take into account rigs that had already been rented or long-term

commitments which the exploitation of a well could necessitate if the oilrig market was tense. We have seen that budgetary estimates concerning exploration were implicitly founded on the principle of total independence of decision making in respect of rigs, but the same independence was obviously not achieved when a single rig was used for several purposes. Historically, this budgetary approach had led to thorny problems when the stand-by phenomenon had to be taken into account, since it was unavoidably linked to all decisions made and not only to an isolated one.
- *The "naval battle"* was one of the standing committee's most useful instruments. All of the subsidiaries' programmes were regularly compared to one another, as well as to the budgetary forecasts which were considered well by well.

An examination of these different techniques shows that they did not make it possible to determine rig-related costs for a specific exploration programme. This type of budget is certainly not easy to evaluate, but its absence indicated the difficulty of integrating the explorer's requirements and those of the rig contractor in the same calculations. Naval, by developing a tool for generating and evaluating programmes, would have to combine in a single instrument these two hitherto unrelated aspects.

None of the above-mentioned techniques, nor the recent creation of special committees, guaranteed that negotiations would produce effective compromises. The oilrig manager was very much aware of the complexity of the projects discussed. He wondered whether it was not possible to develop an instrument which checked the coherence and completeness of information used by the committees, and proposed a programme that reconciled the viewpoints of both head office and subsidiaries. Such a programme could then be submitted to the standing committee and would allow it to pursue discussions and avoid biased negotiations.

It is easy to see why he let himself be convinced by the AI team manager that an expert system could help to draw up such programmes. However, the considerable uncertainties weighing on the project led them to request specific support for the project to clarify unanswered questions, including three salient points. First, if the expert system's objective was a better compromise, who was to be considered as the best expert? Secondly, what place could be left to this type of system in a complex network of actors and committees? Thirdly, how could the system be designed so as to be acceptable to all parties concerned?

3 Reconstruction of automated knowledge: imbalance of expertise

A project group was set up for the development of Naval and placed under the responsibility of the secretary for the oilrig standing committees. At the outset the other members of this group were production counsellors of the relevant geographic area, the team of AI specialists, and the authors in their capacity as participant observers. This project group had to regularly report on its progress to both the rig manager and the AI team manager.

3.1 First steps of the project: who is the planning expert?

One of the project group's first tasks was to define the knowledge and expertise required to build an intelligent system for strategic planning. It also had to identify those experts in the firm required for the development of the project. Three types of knowledge had to be gathered and incorporated into the system:

- *Know-how concerning geology and knowledge of prospects, drilling and the rules of the art in the utilization of oilrigs, as well as the legal conditions to be taken into account in rig contracts.* This knowledge already existed to a large degree and was found throughout the central technical services; it was therefore relatively easy to find representatives of these services who would be able to contribute their experience.
- *Strategic know-how:* many decisions pertaining to the management of oilrigs are guided by strategic considerations. For example, each subsidiary tends to prefer the equipment which it itself has hired, or oilrigs on stand-by are given priority even if there is more suitable or more efficient equipment on the market. In certain cases it is advantageous to rent from a local company, particularly when the subsidiary is involved in negotiations with the government of the host country. These criteria are of prime importance and in the Naval case they came to be seen as decisive when the reasons underlying the different committees' decisions were sought. Even though it was fairly easy to reconstruct and explain the tactical rules which had guided these choices in each of the situations examined, they seemed closely linked to specific contexts which had not been systematically reviewed. In these conditions it was not easy to determine who the best person would be to gather and validate all this knowledge.
- *Planning know-how:* this refers to the methods used to build a programme with the best compromise between conflicting constraints. Very little formalized knowledge on the subject was available in the firm. Programmes were

3 Reconstruction of automated knowledge: imbalance of expertise   207

drawn up by trial and error and elementary knowledge of an essentially temporal nature was used. In contrast, it seemed that use could be made in this field of the promising methods of generating programmes developed by the AI service company. Yet the question of knowing what a good compromise was – and thereby a good programme – remained unanswered.

Finally it appears that what could be called planning expertise in the broad sense, had to combine different types of knowledge in a coherent entity. The project group therefore had to involve experts with each of these categories of expertise, as well as actors capable of combining it all into a single body of knowledge. Paradoxically, the organizational context of the project was to make it easier to mobilize people who had this overall body of knowledge (even though their expertise was neither formally established, nor entirely recognized) rather than the experts of individual disciplines, despite the richness of these disciplines and the ease with which they could be used in the system.

To help the group in identifying the know-how and experts that would contribute to the project, we met with the managers of all the central technical services as well as the representatives of the relevant subsidiaries. Our first observation was that all the actors we met feared another crisis in the near future and felt that the company should prepare for this risk by setting up special procedures. They were, however, also fairly sceptical about the usefulness of a computer system in this respect, since the main problem in a crisis period would be resisting subsidiaries' expansion tendencies. Soaring crude oil prices and euphoric forecasts are always an extremely strong incentive to increase exploration activities and it is in the interests of subsidiaries to reinvest in exploration. At the same time the profitability of prospects declines, which increases the number of candidate prospects. Finally, the scarcity of equipment on the oilrig market forces companies to grasp all available opportunities even if rates are high and the contracts long-term. This analysis has an organizational conclusion: in periods such as these the group's head office has to try and limit rather than encourage its subsidiaries' expansion. But it needs specific actors to fulfill this role and in the Naval case it was hoped that the expert system project would provide them with technical support and expertise. Giving meaning to the concept of a planning team, on both the institutional and technical levels, was therefore one of the stakes and one of the implicit consequences of the project.

In light of such facts one can understand the relative indifference shown by many experts and the problems of involving them in the project group's work. Others, having played a role of informal co-ordination during the stand-by crisis, could however see the stakes implicit in the strengthening of the group's status and resources. Included were notably the secretary of the oilrig standing committee (who was also the leader of the project group) and the regional production counsellor for the relevant geographic zone. During the entire crisis period they had helped the subsidiaries to harmonize their operations programmes with their rig schedules. They had used their knowledge to help draw up programmes, yet without being recognized as planners. Their recent experience encouraged them to become actively involved in the development of Naval.

3.2 Naval: building a programme

After several months the project group was ready to present an early prototype developed by the AI specialists who had designed the system (Figure 1). The latter was divided into two stages: in the first stage Naval tried to identify the most suitable rigs for the exploration programme prospects and in the second stage the system drew up the programme as such. The two stages will now be described.

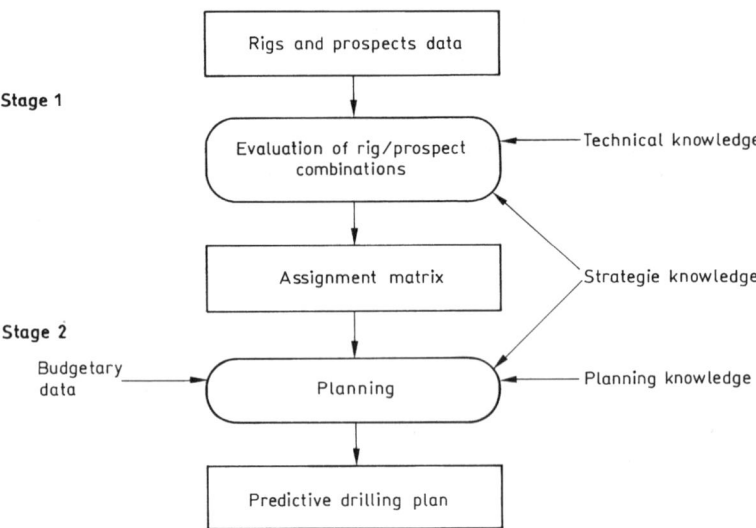

Figure 1 – Architecture of Naval

3.2.1 Evaluating rig-prospect suitability: the value of representations

The data used by the system concerned those prospects which the different subsidiaries had selected for drilling, together with the rigs they were to use – whether these were already rented by the subsidiary or available on the market. These basic objects were described by means of different types of variables. Thus a prospect was defined by its technical characteristics, its duration, the drilling budget, its

3 Reconstruction of automated knowledge: imbalance of expertise

geological value, and its relation to other prospects. A rig was described by its technical capacities, its utilization schedule and its rates and contractual conditions (see Table 1 for a more detailed description of a rig and a prospect).

The aim of this first phase was to evaluate each rig-prospect combination according to the suitability of the rig for drilling that particular prospect. This evaluation was based on rules relating the characteristics of prospects to those of rigs. It had the following structure:

IF (for prospect X, variable I = A)
IF (for rig Y, variable J = B)
THEN the preference given to use of rig Y to drill prospect X = $\alpha \, \varepsilon \, [-1, +1]$.

These preferences varied between -1, which represented total incompatibility (using that rig to drill that prospect was out of the question), and $+1$, which meant perfect compatibility. But preferences of this nature could be determined for each of the twenty or so variables describing rigs and prospects. A list of sometimes contrasting advantages and disadvantages could thus be obtained for each rig-prospect combination. This list had to be combined into a single indicator if the respective advantages of different rigs for drilling a given prospect were to be compared. A multi-criterion function made it possible to integrate these different economic, technical and legal variables.

In spite of its apparent detail, the description of rigs and prospects was in fact fairly rudimentary. It represented only a limited part of the different technical experts' knowledge and thus reflected the difficulties experienced by the project group to involve experts as they were not always convinced of its viability. There was, however, no guarantee that a more sophisticated description of the objects and relations between them would have been useful at that stage of the project. More detail may well have been offset by an unwieldy system and greater difficulties in obtaining a global perception of the properties of rigs and prospects. Several discussions were held in the group on the required degree of detail, a question which was particularly relevant with respect to drilling. It was possible to propose a technical description of rigs and their conditions of use that was just as detailed as the one used by technicians to select a rig from amongst replies to a call for tenders. However, this option was considered too cumbersome and the drillers preferred relying on their thorough knowledge of available rigs and on overall variables – which included their subjective opinion on the advantages and technical capacities of each rig. The terms of the debate differed slightly when it concerned prospects. The decision to drill a prospect did not depend on the geologist's opinion alone; other elements were taken into account, such as the subsidiary director's opinion, contractual obligations vis-à-vis the host State or partners, and technical preparations. Some people considered that Naval provided an opportunity to distinguish these different variables and thereby to facilitate discussions on the advisability of opting for a particular prospect. But there too, only one indicator reflecting the subsidiary's interest in drilling the prospect was finally taken into

Table 1 – Rig and prospect description

Rigs	Prospects
Identification	*Identification of objectives*
Name of rig	Complete name of objective
Contractor's name	Exploration, development, work-over
	Operator's name
General characteristics	Country
Type of rig	
Max. capacity in water depth	*Characteristics of off-shore site*
Min. capacity in water depth	Geographic location of well (in metres)
Type of JackUp	Specific exclusions on the site
Type of sole (JackUp)	Obligation to use a given contract
Weight of derrick (tender)	Existence of a platform
Is the tender self-propelled?	Maximum load on the platform
Drilling capacity	*Technical characteristics of the well*
Diameter of the weakest Bop	Depth (in metres)
Max. pressure of weakest Bop	Maximum pressure at pit-brow
Diameter of the strongest Bop	The presence of H_2S is allowed for
Max. pressure of strongest Bop	Presence of surface gas
Maximum drilling depth	Heavy or light programme
Equipment resistant up to H_2S	Provision for tests
Equipment with a diverter	
Equipment for heavy programmes	*Temporal constraints*
Equipment for light programmes	Start of the time slot
	End of the time slot
Movement characteristics	Expected duration of drilling
knots	
days	*Costs*
	Cost per day excluding rig
Rig value	Current cost per day of an optimum rig
Subjective evaluation of rig's advantages	Rate of group participation
Technical evaluation of rig	
	Evaluation of well
	Planned budgetary year
	Intrinsic value, expectation of discovery
	Progress of preliminary work
	Technical difficulties of operation
	Planning operations
	Last planning decision
	State of progress
	Rig to be used for drilling
	Starting date of drilling
	Values assigned
	To maintaining the well in the budgetary year
	To assigned rig contract
	To maintenance of the planned type of rig
	Flexibility for modifying date

3 Reconstruction of automated knowledge: imbalance of expertise   211

account. At the conclusion of this first stage, the system had all the information it needed to start drawing up a programme:

- budgetary data defining each subsidiary's authorized annual expenses for exploration;
- a description of the characteristics of available rigs and prospects, classified on three levels;
- finally, an allocation matrix, resulting from the evaluation of suitability which made it possible, once a prospect had been chosen, to orientate the choice towards the most suitable rig.

In order to be usable in the planning phase this data was then translated by the system into rules. Each rule represented a suggestion or a constraint which had to be respected, particularly since a high value had been attributed to it.

3.2.2 A compromise algorithm to underpin planning know-how

How was the programme drawn up? The system's reasoning did not follow a sequence of pre-determined steps of the type: "first choose a stand-by rig, then allocate to it as many prospects as it is capable of drilling, then choose another rig and start again…". The application of this kind of algorithm seemed inappropriate since it would have made it necessary from the outset to opt for a single policy (for example either hiring rigs for long periods or signing contracts well by well), whilst the diversity of situations encountered, their complexity and the fluidity of the planning context were unsuited to this type of rigidity. It was more appropriate to find a compromise between the basic characteristics that were ideally to be found in the final programme. These constraints often conflicted with one another, so that the ones which could be respected had to be retained, while those that could not be satisfied without threatening the coherence of the whole had to be eliminated. The quality of a programme depended on the number of contraints which it was able to adhere to. These were expressed in the form of programming rules commonly written as "if A then B with weight factor X".

The following are examples in natural language:

- "a subsidiary's exploration budget must not exceed the allocated budget by more than 10%, with weight factor 9";
- "drilling on prospect A cannot commence before date d, with weight factor 3".

The programme was obtained by the classical method of an arborescent search which used these programming rules and the recommendations resulting from the evaluation phase. The weight factor given to each rule represented the importance given by the user to the corresponding constraint. These weights guided the search for a compromise.

The main steps of the reasoning were as follows. The expert system:

1 – checked that there was a set of partial programmes which respected all the chosen constraints;
2 – chose the constraint with the greatest weight from amongst all those likely to be applied;
3 – tried to apply it, i.e. checked whether there was any contradiction in a completed programme;
4 – resolved any conflict by rejecting a rule accepted in an earlier stage. In backtracking the system always challenged the rule with the lowest weight;
5 – stopped when no more rules could be applied and no conflict was detected;
6 – printed out a GERT system programme which summarized the use of rigs during the next year. It also calculated the budgets of each of these chosen rigs.

The expert system was thus able to build an "optimum" programme by means of the system of weights. Nevertheless, such a programme would be neither "the least expensive" nor "the surest", but simply the best in relation to the advice generated by the rules. In other words, there was no other programme which could have been obtained by rejecting a smaller number of constraints with lower weights.

The formalization of planning expertise gave rise to numerous problems and the number of written rules remained low. Yet an analysis of particularly tricky programmes in the past showed the programme managers' ability to describe how choices had been made, by explaining the different actors' strategies and linking these to changes in the environment. It was the diversity of these situations and their singularity which made the writing of rules so difficult, for the rules had to be simple and few enough to remain operational, yet general enough to apply to a large number of situations. Planning relied more on the tactics applied in a rich and changing context, than on existing expertise. Moreover, the rules which could be formalized appeared somewhat heterogeneous; they concerned expertise or negotiation, knowledge or a convention, they were stable or transient, and yet they were all dealt with in an identical fashion. They therefore contributed towards finding a compromise, since only the weight attributed to them, and not their contents, allowed the system to differentiate them and to give them greater or lesser importance.

This imbalance between the wealth of information on rigs and prospects on the one hand, and the lack of explainable planning knowledge on the other, was characteristic of the context in which the programming of oilrigs took place. To succeed in proposing programmes, Naval had to provide additional external knowledge by incorporating an algorithm of compromise, unknown in former planning methods. How was this compromise algorithm going to behave? Would the proposed programmes be coherent and acceptable?

3 Reconstruction of automated knowledge: imbalance of expertise 213

3.3 A pilot for Naval: a system for experts

In order to provide a quick answer to these questions, the project group embarked on experiments in simple situations. A critical point appeared from the outset, i.e. the preponderant role of the weights in planning. The experts always had trouble in fixing values and justifying them. After several attempts at generating acceptable programmes by using different types of weights, it became clear that there was no point in trying to give these an objective value, for they only had any meaning in relation to one another. They consisted of control variables through which planning strategies could be expressed.

An example will enable us to define more clearly the role and significance of the weight attributed to rules. If, for example, a company wanted to limit the number of wells to be drilled over a period of two budgetary years, so as to spread its activities out evenly, it was possible to write the following kind of rule: "if a goal is programmed in the budget of year n, then over half of the drilling must be realized in that year, with the weight factor X". This rule could have a very different effect, depending on the value given to this weighting. If it had a high value (e.g. $X = 8$), it could cause the choice of certain prospects in this situation to be challenged and to be marked lower than eight. If on the other hand its weight were lower than that assigned to the prospects, it would no longer be able to challenge the choice of a prospect, although it could possibly challenge the choice of a rig.

There was surely no system of weights which could accurately account for the variety of situations experienced by the company. Each specific context could make it necessary to modify the balance of weights to correspond to the most suitable tactics. This need was clearly demonstrated on the occasion of the formalization of several real cases for which the programmes resulting from habitual negotiations were already known. Nevertheless, once the experts had acquired a sound knowledge of both the system's reasoning mechanism and the effect of the weights, they managed in each case to generate a perfectly acceptable programme and, in certain cases and with the system's help, even proposed original or counter-intuitive solutions. The results were encouraging; they confirmed the usefulness of Naval in enhancing the process of generating programmes. On the other hand, using Naval was complex and required a pilot who mastered several types of expertise, with a sound knowledge of rigs and prospects, an understanding of the system's reasoning and the effects of parameters as well as a thorough knowledge of the general planning context, i.e. conditions of the functioning of subsidiaries, state of the rigs, and exploration strategy. The combination of these three areas of expertise was necessary for defining the weights accurately and guiding the compromise, but also for gathering relevant information, explaining the results obtained and justifying the choice of parameters by reducing them to planning logic.

Even though Naval was a knowledge-based system which relied on AI languages and techniques, the term expert system was not altogether accurate in so far as Naval only contained a part of the experts' knowledge. Rather than providing users with new knowledge, it helped them to use the know-how they already had to find an acceptable compromise. Naval was therefore a system for experts rather than an expert system.

4 The project and the development of stakes: a new status for the experts?

4.1 Towards a central planning team

At this stage it was possible to sketch a scenario describing the place and the possible role of the system. It seemed quite natural for it to be made the instrument of experts in a central position in the company; they were in permanent contact with the subsidiaries and the technical departments which provided them with data and informed them of all events concerning rigs and prospects. These experts would be able to draft programmes with the help of the system and propose them to the different committees to use during their negotiations. The programme then decided on by the committees could be submitted to the system to verify its technical coherence or to propose an evaluation of the economic consequences of the choices made.

The system thus called for new organizational developments, something which in itself was hardly surprising since decision making tools often lead to a profound reorganization of their environment. But in this case it required a different way of running the project. The moment seemed ripe for expanding the project group by including new participants from the subsidiaries and the central technical services. The institutional ties which could be established between the central planning team and the other negotiating parties, including the rig committees, had to be defined more clearly. A large number of questions arose. How was recognition of the central planning team's expertise by all the actors concerned to be achieved? How could the gathering of data on the rigs be coordinated with that concerning the prospects? (These two aspects were dealt with separately in the existing organizational set-up.) To what extent would users in the subsidiaries recognize the relevance of programmes proposed and, in particular, the evaluation of a valid programme through a complex weighting system? All these questions were inevitable, they simply meant that it was not possible to envisage the insertion of Naval without reconsidering the exact significance of the planning of drilling activities and without thinking about an organizational logic consistent with this process. In our view, it was possible to start thinking about this without waiting to have a technically faultless instrument. Even if it was still incomplete and im-

4 The project and the development of stakes: a new status for the experts?   215

perfect, the prototype presented the main characteristics of the future system and could already be evaluated.

There was, however, an alternative strategy which consisted of pursuing the work of the project group until a complete system was attained. Indeed, several problems still had to be dealt with, including inadequate user-friendliness, excessive calculation time, insufficient modelling of weights, and so forth. But in our opinion such an alternative had significant drawbacks, for the project would retain its image of a research programme and would not be enhanced by a better definition of its true conditions of utilization. For us, the choice of this second alternative would have been the sign of a bad evaluation of the risks facing the project. These were essentially organizational rather than technical, and nobody could guarantee that a viable and acceptable planning concept would emerge from the project. It was even uncertain whether the existence of a tool was enough to warrant this type of procedure.

We therefore presented these options to the project leaders, for whom the choice was not an easy one. The rig manager preferred the first proposition but found it particularly tricky, for it meant that the project would have to be presented to and accepted by all the actors involved. The leader of the AI group was naturally more inclined to favour the technical development of the system. Discussions on these alternatives were however interrupted by an external event which profoundly affected the company's economic environment. During the autumn of 1985 the price of crude oil plunged by nearly 50%.

4.2 Effects of the counter-shock: the value of planning in a cyclic environment

The fall of crude oil prices totally upset market conditions. Oil companies reduced their exploration programmes and budgets as fast as they could, anticipating a decrease in earnings and reconsidering the evaluation of certain prospects of which the cost-effectiveness was no longer guaranteed. Rig contractors dealt with the depressed market by drastically reducing their rates, but the number of contracted rigs nevertheless continued shrinking throughout 1986.

The policy followed by the oil company directors was fairly simple. They asked their subsidiaries to limit their exploration programmes to their legal obligations and to rent the rigs required for a single well at a time, at the last minute and at the lowest rate. The best strategy was to leave the subsidiaries to act independantly of one another.

Where did Naval fit in, in this type of context? Programmes clearly had a new objective. Co-ordination between subsidiaries no longer aimed at maximizing the use of contracted rigs, at high rates; at this stage its importance was more subtle.

By observing the subsidiaries' decentralized choices, corporate managers wanted to detect the most favourable moment for contracting a rig for a lengthy period at a very low rate. For this operation to be advantageous, they had to be sure of the existence, within a single area, of a programme (possibly grouping together the prospects of several subsidiaries) which could be carried out by the same rig for a period of at least a year or two. Since it was possible to find rigs at very low prices on the depressed market, the financial risks would be fairly small in the event of crude oil prices dropping further, whilst important savings could be generated if the market recovered. Such strategies, by anticipating the appearance of tension on the oilrig market, could place the company in a favourable position in the event of exploration taking off again.

This new conception of planning required the same types of knowledge as that modelled in the first prototype, even if certain rules had to be replaced and the weighting changed. For example, the decision to contract a rig at a low rate for a long period was largely dependent on the quality of the evaluation of prospects in time, and the anticipation of this stability could only be achieved by an in-depth discussion on geological assumptions. Furthermore, a decision on long-term commitments could but come from head office; it had no sense if it concerned an isolated subsidiary. The project henceforth found itself in a strange situation: whereas it had no immediate application in the new environment, it was obvious that with an upturn in the market it would regain its relevance.

At this point it is useful to situate the project's history in the cyclical evolution of its environment. Naval was born at a time when the company was preoccupied with managing the consequences of its commitments, during a period of expanding exploration. Two years later a radically different situation prevailed, with a glut on the oilrig market. This was followed by yet another crisis which inevitably resulted in a restricted market with high rates and the obligation to contract for long periods. Naval, which had originally been designed to assist in the management of this type of tense situation, could be adapted to help anticipate or limit such effects. Pursuing the project implied a long-term vision and a broader planning strategy, capable of mobilizing the expertise and policies adopted at each phase of the economic cycle. But the continuing decline of crude oil prices left no room for the development of such long-term perspectives.

4.3 End of the project: the loss of expertise

As we have seen, the new economic environment seriously threatened the project. If the knowledge used to build Naval was seen as being related to the past only, the project had no use. If it was part of a broader analysis of planning problems and if it was presented as a tool for technical and organizational co-ordination

4 The project and the development of stakes: a new status for the experts?   217

of drilling operations, it maintained its relevance in the long term. Yet the oilrig manager decided to discontinue the project, for two reasons. On the one hand, the new context had affected not only the Naval project but also the entire system of planning drilling operations. The subsidiaries had no difficulty in finding rigs for drilling the rare wells planned in their exploration programmes and they wanted their managerial autonomy. In these conditions, head office had to struggle just to keep the co-ordination committees, set up after 1982, active. On the other hand, financial constraints became necessary at head office and resulted in staff reductions. In particular, the co-ordination and counselling functions assumed by the two main experts had either to be dispensed with, or at least reduced to that of scanning oilrig market trends. Faced with this decentralization trend and the need to reduce costs, the rig manager felt that it was no longer politically feasible to pursue the project, even if it would have been an effective way of preparing for the next crisis.

The AI team tried to see whether the techniques developed during the project could not be used elsewhere. This was perfectly feasible with respect to the principles of finding the best compromise amongst conflicting constraints, but the rest of the programme was too specifically oriented towards the problem of managing drilling operations.

The company was at the same time losing the experience and expertise of the people who had in the past been responsible for the complex task of managing negotiations between head office and the subsidiaries, even if this had been done in an informal manner. Who would be able to limit the number of long-term commitments when the rig market became tense again? There was no simple answer.

To conclude the Naval case history, it seems relevant to look at our support mission and to discuss the role that we were able to play. In spite of our presence, for two years Naval did not develop in what we considered to be the most suitable way. This serves as a reminder that our mission was not a managerial one, and relativizes the significance of our contribution which alone could not ensure the project's success.

Nevertheless, the comparison between a technical project and a monitoring support mission with a broader scope – the project and its environment – is nevertheless useful for two reasons. The first relates to the help provided to the actors so that they do not lose sight of the multiplicity of the objectives. The second reason is more paradoxical and lies in the effort which can be made towards a global understanding of precisely why the project was not what it was expected to be at the time it was evaluated. But by nature innovations follow unexpected courses, and it is not natural for the actors involved in a project with a complex course to question the whys and wherefores of this course.

Clearly, not all corporate projects call for this type of support which only seems useful when a significant margin of uncertainty, even ambiguity, is manifest in the contents, objectives or procedures envisaged. Such was the case with Naval, which in these respects is probably representative of the managerial problems with which firms will be confronted in present and future economic contexts. Naval is also a telling example of the inseparable link between the technical characteristics and the social or economic components of a project. What makes a project evolve is precisely the permanent interaction between these different facets of the same reality, i.e. actors intervening in the modification of their relationships with one another or with their diverse environments (material, legal, financial, etc.).

Bibliography

Aktouf O., *Le management entre tradition et renouvellement*, Québec, Morin, 1989.
Archier G., Serieyx H., *L'entreprise du troisième type*, Paris, Seuil, 1983.
Argyris C., Schön D.A.: *Organizational learning*, Reading, Mass., Addison-Wesley, 1978.
Alter N., *La bureaucratique dans l'entreprise*, Paris, Les Editions Ouvrières, 1985.
Berliner C., Brimson J.A., eds., *Cost management for today's manufacturing*, Cambridge, Mass., Harvard Business School Press, 1989.
Berry M., *La technologie invisible*, Paris, CRG, Ecole Polytechnique, 1985.
Blanc M., Charron E., Freyssenet M., "Le développement des systèmes-experts en entreprise", *Cahiers du GIP*, 35, 1989.
Bonnet A., Haton J.P., Truong-Ngoc J.M., *Systèmes-experts, vers la maîtrise technique*, Paris, InterEditions, 1986.
Bourdaire J.M., Charreton M., *La décision économique*, Paris, PUF, 1989.
Bourgine P., Le Moigne J.L., "Economie de l'intelligence, intelligence de l'économie", *AFCET/Interfaces*, 50, 1986.
Brunsson N., *The irrational organization*, New York, J. Wiley & Son, 1985.
Callon M., "Some elements of a sociology of translation: domestication of the scallops and the fishermen of St. Brieuc", in J. Law, ed., *Power, action and belief*, London, Routledge and Kegan Paul, 1986.
Carlzon J., *La pyramide inversée*, Paris, InterEditions, 1986.
Chandler A.D. Jr., *The visible hand. The managerial revolution in American business*, Cambridge, Mass., Harvard University Press, 1977.
Charue F., "L'apprentissage organisationnel: l'exemple de la robotisation dans l'industrie automobile", Doctoral dissertation, Ecole des Mines de Paris, Paris, 1991.
Chauvet J.M., "Systèmes-experts: la seconde génération", *AFCET/Interfaces*, 53, March, 1987.
Cohen E., *L'Etat brancardier*, Paris, Calmann-Lévy, 1989.
Cohendet P., Hollard M., Malsch T., Veltz P., éds., *L'après-taylorisme*, Paris, Economica, 1988.
Cohendet P., Krasa A., Llerena P., "Principe d'évaluation des processus de production dans un régime de variété permanente", in P. Cohendet, M. Hollard, T. Malsch, P. Veltz, eds., *L'après Taylorisme*, Paris, Economica, 1988.
Crozier M., Friedberg E., *Actors and systems: the politics of collective action*, Chicago, University of Chicago Press, 1980.
David A., Giordano J.L., "Représenter c'est s'organiser", *Gérer et Comprendre*, 22, 1990.
Descottes Y., Latombe J.C., "Compromising among antagonist constraints in a planner", *AI Journal*, 42751: 183-217, 1985.

Dosi G., Freeman C., Nelson R., Silverberg G., Soete L., eds., "Technical change and economic theory", London and New York, Pinter, 1988.

Dreyfus, H.L., Dreyfus S.E., *Mind over machine*, New York, Free Press, 1986.

Dubar C., Dubar E., Engrand S., Feutrie M., Gadrey N., Vermelle M.C., "Innovation de formation et transformation de la socialisation professionelle par et dans l'entreprise", Rapport du Lastrée, Paris, 1989.

Dubar C., *La socialisation. Construction des identités sociales et professionelles*, Paris, Collin, 1991.

Dumond J.P., Heraud D., "La constitution des systèmes-experts et l'appropriation du savoir-faire humain", *Rapport GESTE*, March, 1990.

Farreny H., *Les systèmes-experts: principes et exemples*, Paris, Cepadues Edition, 1986.

Fixari D., Hatchuel A., "Faut-il compléter le modèle japonais?", *Cahier de Recherche*, 2, 1990.

Fixari D., Moisdon J.C., Weil B., *Une expérience originale de requalification d'ouvriers de faible niveau. Rapport de la Mission Nouvelles Qualifications, ANACT*, Paris, Collection Points de Repère, 1991.

Fordyce K., Norden P., Sullivan G., "Expert systems: getting a handle on a moving target", *Interfaces*, 16, 6, 1986.

Goffi J.Y., *Philosophie de la technique*, Paris, PUF, 1988.

Gondran M., *Introduction aux systèmes-experts*, Eyrolles, 1986.

Grundstein M., Bonnières P. de, "De la fertilisation à l'insertion, une méthodologie de développement de système-expert", *Revue d'Intelligence Artificielle*, 2, 1987.

Harmon P., King D., *Expert systems: artificial intelligence in business*, New York, John Wiley & Son, 1985.

Hatchuel A., "L'intervention de chercheurs dans l'entreprise. Elements pour une approche contemporaine", *Revue Education Permanente*, 113, 1988a.

Hatchuel A., "Taylorisme in the age of variety", Paper presented at the meeting "La gestion des entreprises dans une perspective historique", Paris, 1988b.

Hatchuel A., Agrell P., Van Gigch J.P., "Innovation as system intervention", *Systems Research*, 4, 1, 1987.

Hatchuel A., Moisdon J.C., "Théorie de la décision et pratiques organisationnelles. Le cas des investissements pétroliers", *Revue Science et Gestion*, 4, January, 1984.

Hatchuel A., Molet H., "Outils de gestion et logiques de production", Ecole des Mines de Paris, Paris, 1983.

Hatchuel A., Molet H., "Rational modeling in understanding and aiding human decision making", *European Journal of Operational Research*, 24, 1986.

Hatchuel A., Sardas J.C., "Métiers et réseaux: les paradigmes industriels de la gestion de production assistée par ordinateur", *Réseaux*, May-June, 1990.

Heurgon E., ed., *L'avenir de la Recherche Opérationelle*, Paris, Editions Hommes et Téchniques, 1978.

Joseph I., "La relation de service", *Annales de la Recherche Urbaine*, January, 1989.

Latour B., *Science in action: how to follow scientists and engineers through society*, Cambridge, Mass., Harvard University Press, 1987.

Laurent P., "Systèmes-experts et langages orientés objets: un mariage fructueux", *Micro-Systèmes*, July-August, 1987.

Linhart D., *Le torticolis de l'autruche*, Paris, Seuil 1991.

Lorino P., *L'économiste et le manager*, Paris, La Découverte, 1990.
Mac Closkey J.F., "The beginning of operations research: 1934-1941", *Operation Research*, 1, 1987.
Marois T., "Les systèmes-experts: une technique de programmation comme les autres?", *AFCET/Interfaces*, 33, July, 1985.
Masuch P., ed., *Organization, management and expert systems*, Berlin, New York, Walter de Gruyter, 1990.
Maurice M., Sellier F., Silvestre J.J., "The social foundations of industrial power: A comparison of France and Germany", Cambridge, Mass., MIT Press, 1986.
Midler C., "De l'automatisation à la modernisation. Les transformations dans l'industrie automobile", *Gérer et Comprendre*, 13, 1988.
Midler C., *L'auto qui n'existait pas*, Paris, InterEditions, 1993.
Mintzberg H., *The structuring of organizations*, Englewood Cliffs, Prentice-Hall, 1979.
Mintzberg H., *Power in and around organizations*, Englewood Cliffs, Prentice-Hall, 1983.
Moisdon J.C., "Faut-il croire encore à la recherche opérationnelle?", *AFCET/Interfaces*, 1985.
Moisdon J.C., Weil B., "Collective design: lack of communication or shortage of expertise. Analysis of coordination in the development of new vehicles", *Cahier de Recherche* n°3, Paris, 1991.
Montmollin M. de, Pastre O., *Le Taylorisme*, Paris, La Découverte, 1984.
Moutet A., "Ingénieurs et rationalisation en France de la guerre à la crise", in *L'ingénieur dans la société française*, Paris, Les Editions Ouvrières, 1985.
O'Farell P.M., Pingry J., "Expert systems in manufacturing", *Actes des Journées d'Avignon*, June, 1988.
O'Keefe R.M., "Experiences with expert systems in operational research", *Journal of the Operational Research Society*, 7, 1986.
O'Leary D.E., Turban E., "The organizational impact of expert systems", *Human Systems Management*, 7, 1987.
Pave F., *L'illusion informaticienne*, Paris, L'Harmattan, 1989.
Peters T., Waterman R., *In search of excellence*, New York, Harper & Row, 1982.
Pettigrew A.M., "Longitudinal field research on change: theory and practice", *Organization Science*, 1, 3, 1990.
Pfeffer J., *Power in organizations*, Boston, Pitman, 1981.
Poitou J.P., *Le cerveau de l'usine. Histoire des bureaux d'études Renault de l'origine à 1980*, Aix-en-Provence, PUF, 1984.
Plossl G.W., "MRP, yesterday, today and tomorrow", *Production and Inventory Control Review*, Third quarter, 1980.
Pomerol J.C., Levine P., *Systèmes-experts et systèmes d'aides à la décision*, Hermès, Paris, 1989.
Ponssard J.P., *Stratégie industrielle et économie d'entreprise*, New York, McGraw Hill, 1988.
Ponssard J.P., "Genèse de la modélisation en économie d'entreprise", *Laboratoire d'Econométrie*, 318, August, 1989.
Ponssard J.P., Tanguy H., "Planning in firms as an interactive process", *Theory and Decision*, 34, 1993.

Raïffa H., *Decision analysis. Introductory lectures on choice and uncertainty*, Reading, Mass., Addison-Wesley, 1970.
Renouard O., "Evolution des pratiques de planification en raffinerie", *Mémoire de DEA*, Paris, University of Paris IX-Dauphine, 1988.
Riboud A., *Modernisation mode d'emploi*, Paris, Editions 10/18, 1987.
Rose M., *Industrial behaviour*, London, Penguin, 1988.
Roy B., *Méthodologie multicritère d'aide à la décision*, Economica, 1985.
Roy B., "Decision aid and decision making", in C.A. Bana e Costa, ed., *Readings in multiple criteria decision aid*, Berlin, New York, Springer, 1990.
Sainsaulieu R., *L'identité au travail*, Paris, Presses de la fondation nationale des sciences politiques, 1985.
Sainsaulieu R., *Sociologie de l'organisation et de l'entreprise*, Paris, Presses de la Fondation Nationale des Sciences Politiques, 1989.
Sardas J.C., "Credit scoring: les charmes et les dangers d'un outil à la mode", *Gérer et Comprendre*, September, 1986.
Senge P., Morris L., *Learning organization: corporate structure for tomorrow's workplace*, San Francisco, New Leaders, 1993.
Shortliffe E.H., *Computer-based medical consultations: MYCIN*, New York, American Elsevier, 1976.
Simon H.A., *Administrative behavior*, New York, Macmillan, New York, 1947.
Simondon G., *Du mode d'existence des objets techniques*, Paris, Aubier-Montaigne, 1969.
Sombé L., "Raisonnements sur des informations incomplètes en intelligence artificielle", *Revue d'Intelligence Artificielle*, 3-4, 1988.
Stamper R., "Pathologies of AI: responsible use of artificial intelligence in professional work." *AI & Society*, 2, 1988.
Tanguy H., "La réhabilitation des modèles et des plans: le cas d'une maison de Champagne", *Cahiers d'économie et de sociologie rurale*, 10, 1989.
Taylor F.W., *The principles of scientific management*, New York, Harper & Brothers, 1911.
Thompson J.D., *Organizations in action*, New York, McGraw Hill, 1967.
Tsang J.P., *Planification par combinaison de plans; application à la génération de gommes d'usinage*, Doctoral dissertation, Grenoble, University of Grenoble, 1987.
Van Der Gaag L., Lucas P., "An overview of expert system principles", in "*Organization, management and expert-systems*", Masuch M., ed., Berlin, New York, Walter de Gruyter, 1990.
Vernant J.P., *Myth and thought among the Greeks*, New York, Routledge, 1983.
Walton R.E., Susman G.I., "People policies for the new machines", *Harvard Business Review*, March-April, 1987.
Whittaker D.H., *Managing innovation. A study of British and Japanese factories*, Cambridge, UK, Cambridge University Press, 1990.